THE MORAL CASE FOR FOSSIL FUELS

THE MORAL CASE FOR FOSSIL FUELS

ALEX EPSTEIN

PORTFOLIO / PENGUIN

PORTFOLIO / PENGUIN
Published by the Penguin Group
Penguin Group (USA) LLC
375 Hudson Street
New York, New York 10014

USA | Canada | UK | Ireland | Australia | New Zealand | India | South Africa | China
penguin.com
A Penguin Random House Company

First published by Portfolio / Penguin, a member of Penguin Group (USA) LLC, 2014

Image on page 115: Craig D. Idso, Center for the Study of Carbon Dioxide
and Global Change.
Other images courtesy of the author

ISBN 978-1-59184-744-1

Printed in the United States of America
10 9 8 7 6 5 4 3 2 1

Set in ITC New Baskerville Std
Designed by Elyse Strongin

CONTENTS

1

THE SECRET HISTORY OF FOSSIL FUELS

"YOU MUST MAKE A LOT OF MONEY"

"You're an environmentalist, right?" the girl, college age, asked me. It was 2009, in Irvine, California. I had stopped at a farmers' market near my office for lunch, and she was manning a Greenpeace booth right next to it.

"Do you want to help us end our addiction to dirty fossil fuels and use clean, renewable energy instead?"

"Actually," I replied, "I study energy for a living—and I think it's good that we use a lot of fossil fuels. I think the world would be a much better place if people used a lot more."

I was curious to see how she would respond—I doubted she had ever met anyone who believed we should use *more* fossil fuels. I was hoping that she would bring up one of the popular arguments for dramatically reducing fossil fuel use, and I could share with her why I thought the benefits of using fossil fuels far outweighed the risks.

But fossil fuels cause climate change, she might have said. I agree, I would have replied, but I think the evidence shows that climate change, natural or man-made, is more manageable than ever, because human beings are so good at adapting, using ingenuity and technology.

But fossil fuels cause pollution, she might have said. I agree, I would have replied, but I think the evidence shows that ingenuity and technology make pollution a smaller problem every year.

But fossil fuels are nonrenewable, she might have said. I agree, I would have replied, but I think the evidence shows that there are huge amounts of fossil fuels left, and we'll have plenty of time to use ingenuity and technology to find something cheaper—such as some form of advanced nuclear power.

But fossil fuels are replaceable by solar and wind, she might have said. I disagree, I would have replied, because the sun and the wind are intermittent, unreliable fuels that always need backup from a reliable source of energy—usually fossil fuels, which is the only source of energy that has been able to provide cheap, plentiful, reliable energy for the billions of people whose lives depend on it.

But she didn't say any of those things. Instead, when I said I thought that we should use more fossil fuels, she looked at me with wide-eyed disbelief and said, "Wow, you must make a lot of money."

In other words, the only conceivable reason I would say that our use of fossil fuels is a good thing is if I had been paid off by the fossil fuel industry.

Even though this wasn't true, I understood why she thought it. It is conventional wisdom that our use of fossil fuels is an "addiction"—a short-range, unsustainable, destructive habit.

Eighty-seven percent of the energy mankind uses every second, including most of the energy I am using as I write this, comes from burning one of the fossil fuels: coal, oil, or natural gas.[1] Every time someone uses a machine—whether the computer I am using right now, the factory it was produced in, the trucks and ships that transported it, the furnace that forged the aluminum, the farm equipment that fed all the workers who made it, or the electricity that keeps their lights on, their phones charged, and their restaurants and hospitals open—they are using energy that they must be able to rely on and afford. And 87 percent of the time, that energy comes from coal, oil, or natural gas.[2] Without exception, anyone who lives a modern life is directly or indirectly using large amounts of fossil fuel energy—it is that ubiquitous.

But, we are told, this cannot continue.

While it might be convenient to drive gasoline cars and get electricity from coal in the short run, and while we might have needed them in the past, the argument goes, in the long run we are making our climate unlivable, destroying our environment, and depleting our resources. We must and can replace fossil fuels with renewable, green, climate-friendly energy from solar, wind, and biomass (plants).

This is not a liberal view or a conservative view; it's a view that almost everyone holds in one form or another. Even fossil fuel companies make statements like the one the former CEO of Shell made in 2013: "We believe climate change is real and time is running out to take real action to reduce greenhouse gas emissions."[3] President George W. Bush was the person who popularized the expression "addicted to oil."[4] The debate over our addiction to fossil fuels is usually over *how dangerous* the addiction is and *how quickly* we can get rid of it—not whether we have one.

And the most prominent groups say we must get rid of it very quickly.

For years, the Nobel Prize–winning Intergovernmental Panel on Climate Change (IPCC) has demanded that the United States and other industrialized countries cut carbon dioxide emissions to 20 percent of 1990 levels by 2050—and the United States has joined hundreds of other countries in agreeing to this goal.[5]

Every day, we hear of new predictions from prestigious experts reinforcing the calls for massive restrictions on fossil fuel use. As

I write this, news about melting ice in West Antarctica is leading to dire predictions of sea level rises: "Scientists Warn of Rising Oceans from Polar Melt," reports the *New York Times*; "Is It Too Late to Save Our Cities from Sea-Level Rise?" asks *Newsweek*, citing new research that "Miami and Manhattan will drown sooner than we thought."[6]

The message is clear: Our use of fossil fuels is going to destroy us in the long run, and we should focus our efforts on dramatically reducing it as soon as humanly possible.

So when the girl at the Greenpeace booth implied that I had sold my soul, I didn't get offended. I simply explained that, no, I wasn't being paid off; I had just concluded, based on my research, that the short- and long-term benefits of using fossil fuels actually far, far outweigh the risks and was happy to explain why. But she wasn't interested. Pointing me to the Greenpeace pamphlets giving all the reasons fossil fuels are bad, she said, "So many experts predict that using fossil fuels is going to lead to catastrophe—why should I listen to you?" She made it clear that this wasn't a real question and that the conversation was over.

But if she had wanted an answer, I would have told her this: I understand that a lot of smart people are predicting catastrophic consequences from using fossil fuels, I take that very seriously, and I have studied their predictions extensively.

And what I have found is this: leading experts and the media have been making the exact same predictions for more than thirty years. As far back as the 1970s they predicted that if we did not dramatically reduce fossil fuel use *then*, and use renewables instead, we would be experiencing catastrophe *today*—catastrophic resource depletion, catastrophic pollution, and catastrophic climate change. Instead, the exact opposite happened. Instead of using a lot less fossil fuel energy, we used a lot more—but instead of long-term catastrophe, we have experienced dramatic, long-term improvement in every aspect of life, including environmental quality. The risks and side effects of using fossil fuels declined while the benefits—cheap, reliable energy and everything it brings—expanded to billions more people.

This is the secret history of fossil fuels. It changed the way I think about fossil fuels and it may change the way you think about them, too.

DÉJÀ VU

When I was twenty years old, I decided I wanted to write about "practical philosophy" for a living. Philosophy is the study of the basic principles of clear thinking and moral action. While college philosophy classes all too often present philosophy as an impractical subject that involves endlessly debating skeptical questions ("How do you know you exist?" "How do you know you're not in The Matrix?"), philosophy is in fact an incredibly practical tool. No matter what we're doing in life, whether we're coming up with a business plan or raising children or deciding what to do about fossil fuels, it is always valuable to be able to think clearly about what is right and what is wrong and why.

One valuable lesson philosophy taught me is that with any idea, such as the idea that we need to get off fossil fuels, we should look at the *track record* of that idea, if it has one.

Now, you might think: this idea does not have a history because it is a *new* idea based on the latest science. This is certainly the impression many of our leading intellectuals give. For example, in 2012 I debated Bill McKibben, the world's leading opponent of fossil fuels, at Duke University, and he presented his view of our addiction to fossil fuels as cutting-edge: "We should be grateful for the role that fossil fuel played in creating our world and equally grateful that scientists now give us ample warning of its new risks, and engineers increasingly provide us with the alternatives that we need."[7] This is the narrative we hear over and over: fossil fuels were once necessary, but the latest science tells us they're causing an imminent catastrophe unless we stop using them and replace them with cutting-edge renewables.

What is rarely mentioned is that thirty years ago, leading experts, including many of today's leading experts, were telling us that fossil fuels were once necessary, but the latest science tells us they're causing an imminent catastrophe unless we stop using them and replace them with cutting-edge renewables.

Take the prediction we hear today that we will soon run out of fossil fuels—particularly oil—because they are nonrenewable. This prediction was made over and over by some of the most prestigious thinkers of the 1970s, who assured us that their predictions were backed by the best science.

In 1972, the international think tank the Club of Rome released a multimillion-copy-selling book, *The Limits to Growth,* which declared that its state-of-the-art computer models had demonstrated that we would run out of oil by 1992 and natural gas by 1993 (and, for good measure, gold, mercury, silver, tin, zinc, and lead by 1993 at the latest).[8] The leading resource theorist of the time was ecologist Paul Ehrlich, who was so popular and prestigious that Johnny Carson invited him onto his show over a dozen times. In 1971 he said, "By the year 2000 the United Kingdom will be simply a small group of impoverished islands, inhabited by some 70 million hungry people,"[9] and in 1974 he wrote, "America's economic joyride is coming to an end: there will be no more cheap, abundant energy, no more cheap abundant food."[10]

Another catastrophic prediction we hear today is that pollution from fossil fuels will make our environment more and more hazardous to our health—hence we need to stop using "dirty" fossil fuels. This prediction was also made many times in the 1970s—with many assurances that these predictions were backed by the best science.

Life magazine reported in January 1970 that, because of particles emitted in the air by burning fossil fuels, "Scientists have solid experimental and theoretical evidence to support . . . the following predictions: In a decade, urban dwellers will have to wear gas masks to survive air pollution . . . by 1985 air pollution will have reduced the amount of sunlight reaching earth by one half . . ."[11] To quote Paul Ehrlich again, as he may have been the most influential public intellectual of the decade (and is still a prestigious professor of ecology at Stanford University): "Air pollution . . . is certainly going to take hundreds of thousands of lives in the next few years alone," he said in 1970.[12]

And then there's the prediction we hear most today: the supposedly scientifically indisputable claim that CO_2 emissions from fossil fuels will cause a true climate catastrophe within a couple of decades.[13] Reading back in time, I saw that many of the leaders who make that prediction now had, decades ago, predicted that we'd be living in catastrophe *today*.

Here's a 1986 news story about a prediction by James Hansen, the most influential climate scientist in the world over the last thirty years:

> Dr. James E. Hansen of the Goddard Space Flight Center's Institute for Space Studies said research by his institute showed that because of the "greenhouse effect" that results when gases prevent heat from escaping the earth's atmosphere, global temperatures would rise early in the next century to "well above any level experienced in the past 100,000 years."
>
> Average global temperatures would rise by one-half a degree to one degree Fahrenheit from 1990 to 2000 if current trends are unchanged, according to Dr. Hansen's findings. Dr. Hansen said the global temperature would rise by another 2 to 4 degrees in the following decade.[14]

Bill McKibben, when he told Duke students in 2012 that we were on the verge of drastic warming, neglected to mention the results of his decades-old claims, such as this one in 1989: "The choice of doing nothing—of continuing to burn ever more oil and coal—is not a choice, in other words. It will lead us, if not straight to hell, then straight to a place with a similar temperature"; and "a few more decades of ungoverned fossil-fuel use and we burn up, to put it bluntly."[15]

John Holdren, a protégé of Paul Ehrlich who serves as science adviser to President Barack Obama, had a particularly dire prediction, according to Ehrlich in 1986: "As University of California physicist John Holdren has said, it is possible that carbon-dioxide climate-induced famines could kill as many as a billion people before the year 2020."[16]

Just as the media today tell us these catastrophic predictions are a matter of scientific consensus, so did the media of the 1980s. For example, on the issue of catastrophic climate change: "By early 1989 the popular media were declaring that 'all scientists' agreed that warming was real and catastrophic in its potential," a 1992 study reported.[17]

If all the predicted catastrophes—depletion, pollution, climate change—had occurred as thought leaders said they would, the world of today would be much, much worse than the world of the 1970s. In the 1970s, Ehrlich went as far as to say, of the overall devastation ahead, "If I were a gambler, I would take even money that England will not exist in the year 2000."[18]

And these were not idle predictions—the coming fossil fuel catastrophe was so bad, these leading experts said, that we needed dramatic restrictions on fossil fuel energy use. Ehrlich wrote: "Except in special circumstances, all construction of power generating facilities should cease immediately, and power companies should be forbidden to encourage people to use more power. Power is much too cheap. It should certainly be made more expensive and perhaps rationed, in order to reduce its frivolous use."[19]

In 1977, Amory Lovins, widely considered the leading energy thinker of the 1970s for his criticisms of fossil fuels and nuclear power and his support of solar power and reduced energy use, explained that we already used too much energy. And in particular, the kind of energy we least needed was . . . *electricity*, the foundation of the digital/information revolution: "[W]e don't need any more big electric generating stations. We already have about twice as much electricity as we can use to advantage."[20]

In 1998, Bill McKibben endorsed a scenario of outlawing 60 percent of present fossil fuel use to slow catastrophic climate change, even though that would mean, in his words, that "each human being would get to produce 1.69 metric tons of carbon dioxide annually—which would allow you to drive an average American car nine miles a day. By the time the population increased to 8.5 billion, in about 2025, you'd be down to six miles a day. If you carpooled, you'd have about three pounds of CO_2 left in your daily ration—enough to run a highly efficient refrigerator. Forget your computer, your TV, your stereo, your stove, your dishwasher, your water heater, your microwave, your water pump, your clock. Forget your light bulbs, compact fluorescent or not."[21]

All of these thinkers still advocate similar policies today—in fact, today Bill McKibben endorses a *95 percent ban on fossil fuel* use, *eight times* as severe as the scenario described above![22] And all of them are extremely prestigious. Since making these predictions, John Holdren has become science adviser to President Obama, Bill McKibben is called "the nation's leading environmentalist"[23] and more than anyone led opposition to the Keystone XL pipeline, and Paul Ehrlich is still arguably the most influential ecological thinker in the world. Energy historian Robert Bradley Jr. chronicles his accolades:

> Ehrlich held an endowed chair as the Bing Professor of Population Studies in the Biology Department at Stanford and was elected president of the American Institute of Biological Sciences. He was elected to the National Academy of Sciences and received many awards and prizes, including the inaugural prize of the American Academy of Arts and Sciences for Science in the Service of Humanity, a MacArthur Genius Award, the Volvo Environmental Prize, the World Ecology Medal from the International Center for Tropical Ecology, and the International Ecology Institute Prize.
>
> He also received what is hyped as the equivalent of the Nobel Prize in a field where it is not awarded—the Crafoord Prize in Population Biology and the Conservation of Biological Diversity.[24]

Thus, today's leading thinkers and leading ideas about fossil fuels have a decades-long track record—and given that they are calling for the abolition of our most popular form of energy, it would be irresponsible not to look at how reality has compared to their predictions.

Of course, predictions on a societal or global level can never be exact, but they need to be somewhere near the truth.

So what happened?

Two things: Instead of following the leading advice and restricting the use of fossil fuels, people around the world nearly doubled their use of fossil fuels—which allegedly should have led to an epic disaster. Rather, it led to an epic improvement in human life across the board.

MORE FOSSIL FUELS, MORE FLOURISHING

Here is a picture summarizing world energy use since 1980.

Figure 1.1: 80 Percent Increase in Worldwide Fossil Fuel Use 1980–2012

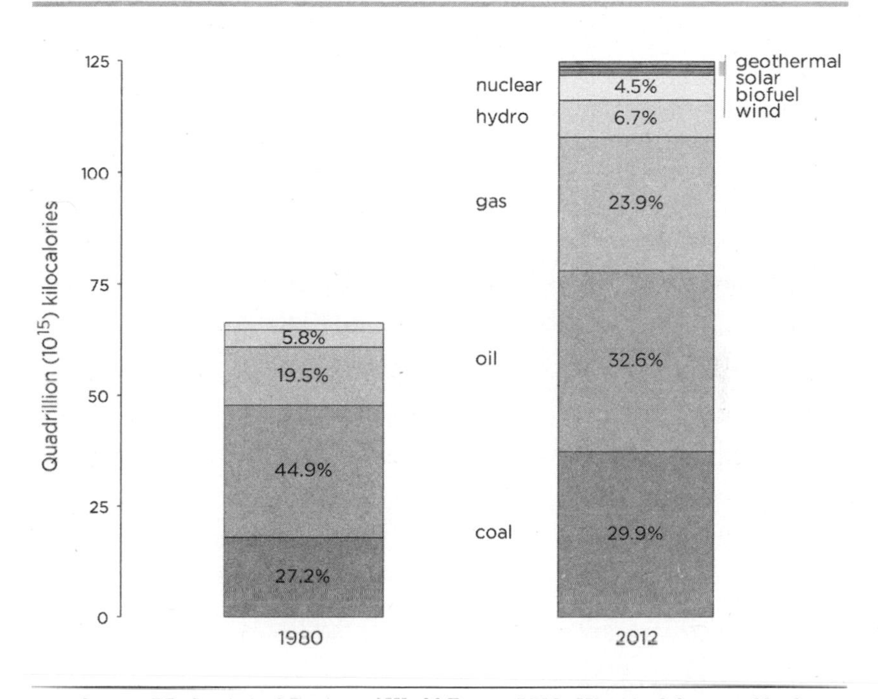

Source: BP, Statistical Review of World Energy 2013, Historical data workbook

From the 1970s to the present, fossil fuels have overwhelmingly been the fuel of choice, particularly for developing countries. In the United States between 1980 and 2012, the consumption of oil increased 8.7 percent, the consumption of natural gas increased 28.3 percent, and the consumption of coal increased 12.6 percent.[25] During that time period, the world overall increased fossil fuel usage far more than we did. Today the world uses 39 percent more oil, 107 percent more coal, and 131 percent more natural gas than it did in 1980.[26]

This wasn't supposed to happen.

The anti–fossil fuel experts had predicted that this would be not only deadly, but unnecessary due to the cutting-edge promise of solar and wind (sound familiar?). Then as now, environmental leaders were arguing that renewable energy combined with conservation—using less energy—was a viable replacement for fossil fuels.

Amory Lovins wrote in 1976: "Recent research suggests that a largely or wholly solar economy can be constructed in the United States with straightforward soft technologies that are now demonstrated and now economic or nearly economic."[27] Lovins was a sensation, and around the globe governments gave solar (and wind and ethanol) companies billions of dollars in the hope that they would be able to generate cheap, plentiful, reliable energy.

But as the last graph illustrates, this did not happen. Solar and wind are a minuscule portion of world energy use. And even that is misleading because fossil fuel energy is reliable whereas solar and wind aren't. While energy from, say, coal is available on demand so you can keep a refrigerator—or a respirator—on whenever you need it, solar energy is available only when the sun shines and the clouds cooperate, which means it can work only if it's combined with a reliable source of energy, such as coal, gas, nuclear, or hydro.[28]

Why did fossil fuel energy outcompete renewable energy—not just for existing energy production but for most new energy production? This trend is too consistent across too many countries to be ignored. The answer is simply that renewable energy couldn't meet those countries' energy needs, though fossil fuels could. While many countries wanted solar and wind, and in fact used a lot of their citizens' money to prop up solar and wind companies, no one could figure out a cost-effective, scalable *process* to take sunlight and wind, which are dilute and intermittent forms of energy, and turn them into cheap, plentiful, reliable energy.

So despite the warnings of leading experts, people around the world nearly doubled their use of fossil fuels.

According to the predictions of the most popular experts, who assured us that their conclusions reflected the best science, this should have led to utter catastrophe. But the result was one of the greatest-ever improvements in human life.

This book is about morality, about right and wrong. To me, the question of what to do about fossil fuels and any other moral issue comes down to: What will promote human life? What will promote human *flourishing*—realizing the full potential of life? Colloquially, how do we maximize the years in our life and the life in our years? When we look at the recent past, the past that was supposed to be so disastrous, we should look at flourishing—and that of course includes the quality (or lack thereof) of our environment.

And there is an incredibly strong correlation between fossil fuel use and life expectancy and between fossil fuel use and income, particularly in the rapidly developing parts of the world. Figures 1.2 and 1.3 show recent trends in China and India of fossil fuel use, life expectancy, and income.

There is no perfect measure of flourishing, but one really good measure is life expectancy—the average number of years in the life of a human being. Another good one, for less obvious reasons, is average income. This is valuable because while in a sense "money can't buy happiness," it gives us *resources* and, therefore, time and opportunity to pursue our happiness. It's hard to be happy when you don't know where your next meal is coming from. The more opportunity you have to do what you want with your time, the more opportunity you have to be happy.

Consider the fate of two countries that have been responsible for a great deal of the increase in fossil fuel use, China and India. In each country, both coal and oil use increased by *at least a factor of 5*, producing nearly all their energy.[29]

Figure 1.2: Fossil Fuel Use and Life Expectancy in China and India

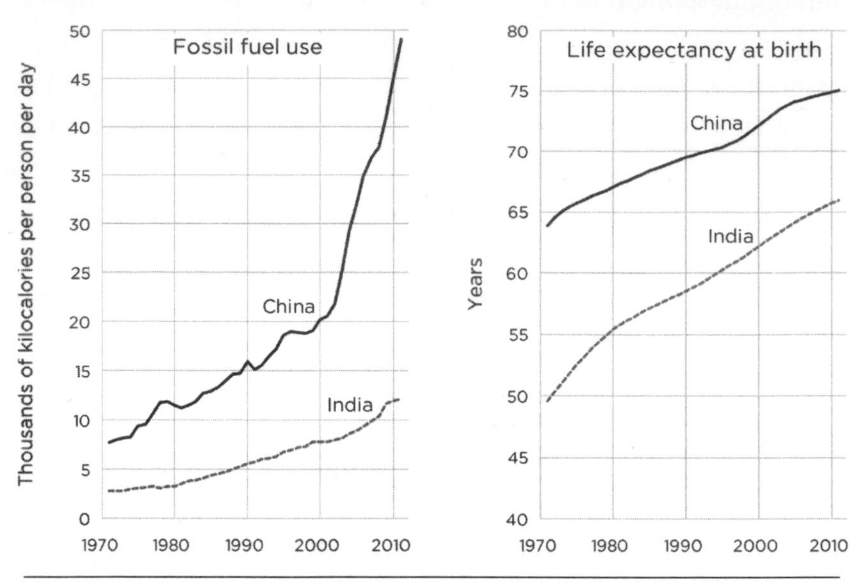

Sources: BP, Statistical Review of World Energy 2013, Historical data workbook; World Bank, World Development Indicators (WDI) Online Data, April 2014

Figure 1.3: Fossil Fuel Use and Income in China and India

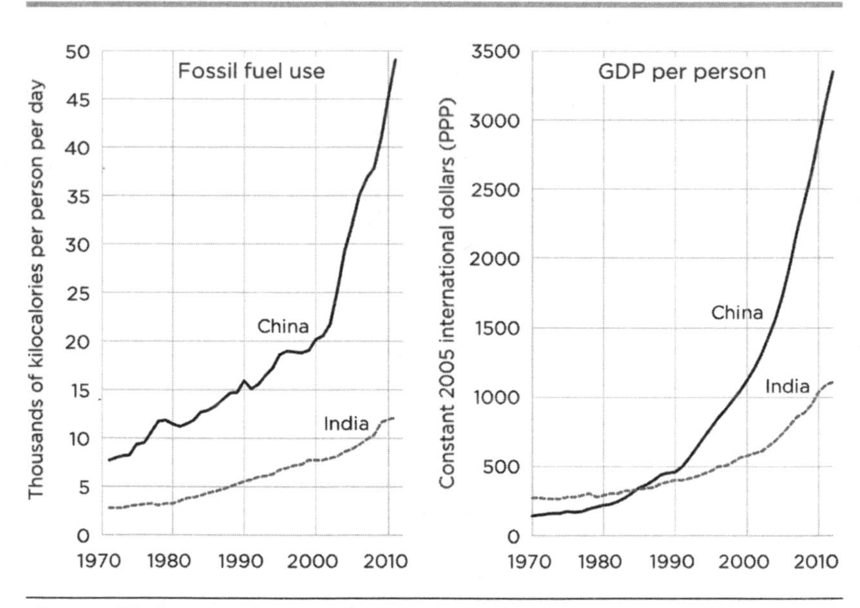

Sources: BP, Statistical Review of World Energy 2013, Historical data workbook; World Bank, World Development Indicators (WDI) Online Data, April 2014

The story is clear—both life expectancy and income increased rapidly, meaning that life got better for billions of people in just a few decades. For example, the infant mortality rate has plummeted in both countries—in China by 70 percent, which translates to 66 more children living per 1000 births.[30] India has experienced a similar decrease, of 58 percent.

Not only in China and India, but around the world, hundreds of millions of individuals in industrializing countries have gotten their first lightbulb, their first refrigerator, their first decent-paying job, their first year with clean drinking water or a full stomach. To take one particularly wonderful statistic, global malnutrition and undernourishment have plummeted—by 39 percent and 40 percent, respectively, since 1990.[31] That means, in a world with a growing population, billions of people are better fed than they would have been just a few decades ago. While there is plenty to criticize in how certain governments have handled industrialization, the big-picture effect has been amazingly positive so far.

Ours is a world that was not supposed to be possible.

Where did the thinkers go wrong? One thing I have noticed in reading most predictions of doom is that the "experts" almost always focus on the *risks* of a technology but never the benefits—and on top of that, those who predict the most risk get the most attention from the media and from politicians who want to "do something."

But there is little to no focus on the *benefits* of cheap, reliable energy from fossil fuels.

This is a failure to think big picture, to consider *all* the benefits and *all* the risks. And the benefits of cheap, reliable energy to power the machines that civilization runs on are enormous. They are just as fundamental to life as food, clothing, shelter, and medical care—indeed, all of these require cheap, reliable energy. By failing to consider the benefits of fossil fuel energy, the experts didn't anticipate the spectacular benefits that energy brought about in the last thirty years.

At the same time, we do have to consider the risks—including predictions that using fossil fuel energy will lead to catastrophic resource depletion, catastrophic pollution, and catastrophic climate change.

How did those predictions fare? Even if the overall trends are positive, might the anti–fossil fuel experts have been right about catastrophic depletion, catastrophic pollution, and catastrophic climate change, and might those problems still be leading us to long-term catastrophe?

These are important questions to answer.

But when we look at the data, a fascinating fact emerges: As we have used more fossil fuels, our resource situation, our environment situation, and our climate situation have been improving, too.

MORE FOSSIL FUELS, MORE RESOURCES, BETTER ENVIRONMENT, SAFER CLIMATE?

Let's start with the popular prediction that we're running out of resources, especially fossil fuels.

If the predictions were right that we were running out of fossil fuel resources, then nearly doubling fossil fuel use worldwide should have practically depleted us of fossil fuels, even faster than Paul Ehrlich and others predicted. That's certainly what the experts told us in the 1970s. In a 1977 televised address, Jimmy Carter, conveying conventional wisdom at the time, told the nation, "We could use up all of the proven reserves of oil in the entire world by the end of the next decade."[32] A popular Saudi expression at the time captured this idea: "My father rode a camel. I drive a car. My son flies a jet airplane. His son will ride a camel."[33]

Well, no one in the oil business is riding a camel, because as fossil fuel *use* has increased, fossil fuel *resources* have increased. How is that possible?

The measure for fossil fuel resources is "proven reserves," which is the amount of coal, oil, or gas that is available to us affordably, given today's technology. While these statistics are subject to some manipulation—sometimes countries and companies can give misleading data—they are the best information we have and we have historically *underpredicted* availability.

Let's look at reserves from 1980 to the present for oil and gas, the fossil fuels we are traditionally afraid will run out. Coal is much easier to find and extract and is considered to be the fossil fuel that is least likely to run out. Notice how the more we consume, the more reserves increase.

Figure 1.4: More Oil Consumption, More Oil Reserves

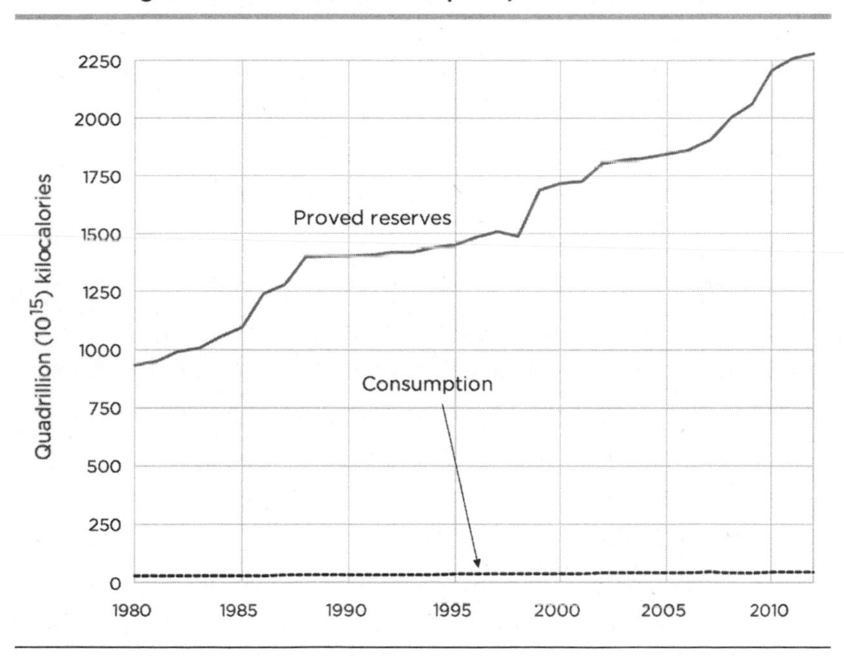

Source: BP, Statistical Review of World Energy 2013, Historical data workbook

Figure 1.5: More Natural Gas Consumption, More Natural Gas Reserves

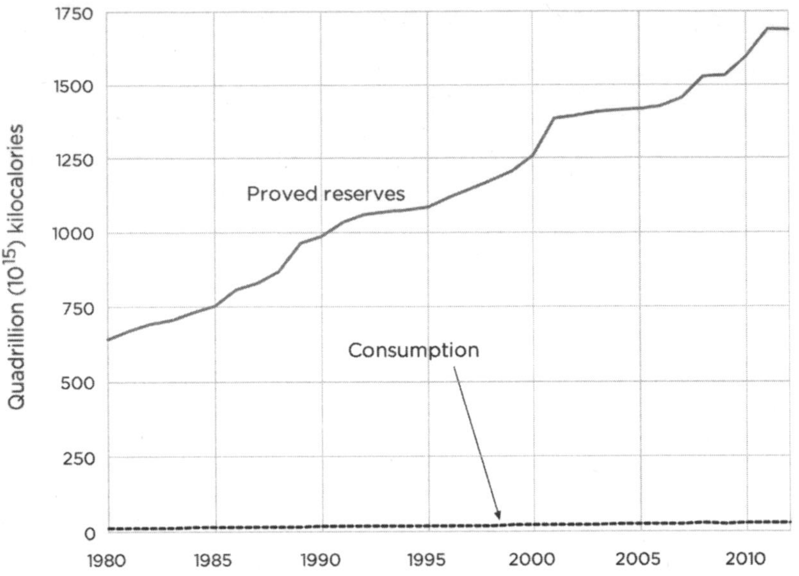

Source: BP, Statistical Review of World Energy 2013, Historical data workbook

This is counterintuitive; the more we use, the more we have.

How did this happen? Stay tuned.

Why did so many expect catastrophic depletion? Again, there was a failure to think big picture. Many experts paid attention only to our consumption of oil and gas resources, but not our ability to create new oil and gas resources.

It's true that once we burn a barrel of oil, it's gone. But it's also true that *human ingenuity* can dramatically increase the amount of coal, oil, or gas that is available. It turns out that there are many times more of each in the ground than we have used in the entire history of civilization—it's just a matter of developing the technology to extract them economically.[34] And in general, human beings are amazingly good at using ingenuity to create wealth, which means to create resources. We take the materials around us and make them more valuable; that's how we went from starving in a cave to producing a cornucopia of food that we can enjoy in comfortable homes. The thought leaders did not sufficiently consider these virtues of human beings.

What about the prediction that our environment would degrade as we used more fossil fuels and more everything? Our escalating fossil fuel use was definitely supposed to be punished with a much, much dirtier environment.

What actually happened? We'll look at all major measures of environmental quality in chapter 8, but for now let's look at clean air and clean water. Both have increased substantially.

Here are measurements from the EPA of six major air pollutants. As fossil fuel use goes up, they go down.

Figure 1.6: U.S. Air Pollution Goes Down Despite Increasing Fossil Fuel Use

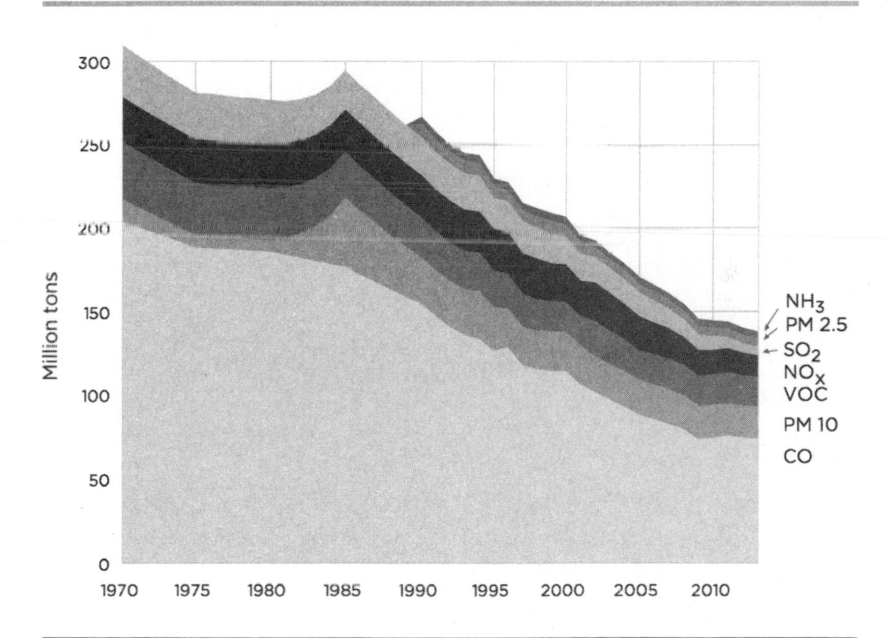

Source: U.S. EPA National Emissions Inventory Air Pollutant Emissions Trends Data

And here are international data for the percentage of people in the world with good water quality, which has gone up dramatically in the last 25 years as countries have used more and more fossil fuels.

Figure 1.7: More Fossil Fuels, More Clean Water

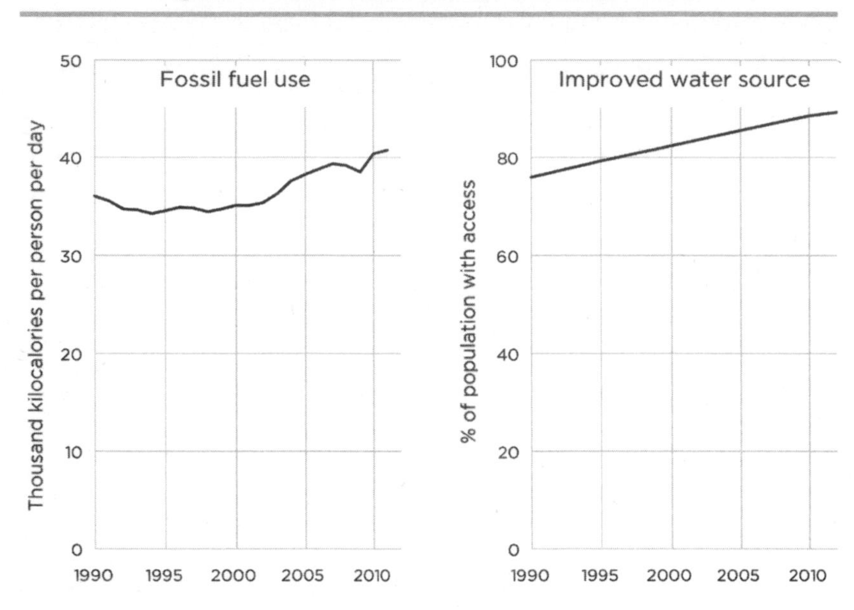

Sources: BP, Statistical Review of World Energy 2013, Historical data workbook; World Bank, World Development Indicators (WDI) Online Data, April 2014

Overall, the improvement is incredible. Of course, there are places such as China that have high levels of smog—but the track record of the rest of the world indicates that this can be corrected while using ever increasing amounts of fossil fuels.

Once again, the anti–fossil fuel experts got it completely wrong. Why?

Again by not thinking big picture, by paying attention to only one half of the equation—in the case of fossil fuels, focusing only on the ways in which using them can harm our environment. But fossil fuels, as we'll discuss in chapter 6, can also *improve* our environment by powering machines that clean up nature's health hazards, such as water purification plants that protect us from naturally contaminated water and sanitation systems that protect us from natural disease and animal waste. Pessimistic predictions often assume that our environment is perfect until humans mess it up; they don't consider the possibility that we could improve our environment. But the data of the last forty years indicate that we have been doing exactly that—using fossil fuels.

Finally, we have to look at what the trend is in the realm of climate change. Catastrophic climate change is the most dire claim about fossil fuels today, and it is associated with many prominent scientific bodies, journals, and media outlets—although if we go through the writings of the 1970s and 1980s, we see those same bodies declare many things confidently about *global cooling* only to contradict themselves several years later. In 1975, the American Meteorological Society told Americans that the climate was cooling and that this meant worse weather: "Regardless of long term trends, such as the return of an Ice Age, unsettled weather conditions now appear more likely than those of the abnormally favorable period which ended in 1972."[35] In 1975, *Nature* said, "A recent flurry of papers has provided further evidence for the belief that the Earth is cooling. There now seems little doubt that changes over the past few years are more than a minor statistical fluctuation."[36]

In the late 1970s, the global cooling trend many expected to end in disaster ended with no disaster whatsoever.

Since then, those who believe in catastrophic climate change have overwhelmingly focused on global warming due to CO_2 emissions from fossil fuels. It has long been known that when CO_2 is added to the atmosphere, the greenhouse effect leads to a warming impact—but before the 1970s and 1980s, there was not much fear that it was of a significant enough magnitude to do major harm (or good, for that matter). But starting in the 1970s and especially the

1980s, claims of runaway global warming and resulting catastrophic climate change became popular. How did they fare when compared to reality?

Recall that in 1986 James Hansen predicted that "if current trends are unchanged," temperatures would rise .5 to 1.0 degree Fahrenheit in the 1990s and 2 to 4 degrees in the first decade of the 2000s.[37] According to Hansen's own department at NASA, from the beginning to the end of the 1990s, temperatures were .018 degree Fahrenheit (.01 degree Celsius) higher, and from 2000 to 2010, temperatures were .27 degree Fahrenheit (.15 degree Celsius) higher—meaning he was wrong many times over.[38]

Recall also that journalist Bill McKibben, summarizing the claims of Hansen and others, confidently predicted that by now we would "burn up, to put it bluntly."[39] Looking at the actual data on a graph, it becomes clear that he was completely wrong.

Here's a graph of the last hundred-plus years of temperature compared to the amount of CO_2 in the atmosphere. We can see that CO_2 emissions rose rapidly, most rapidly in the last fifteen years. But there is not nearly the warming or the pattern of warming that we have been led to expect. We can see a very mild warming trend overall—less than 1 degree Celsius (less than 1.5 degrees Fahrenheit) over a century—which in itself is unremarkable, given that there is always a trend one way or the other, depending on the time scale you select. But notice that there are smaller trends of warming and cooling, signifying that CO_2 is not a particularly powerful driver, and especially notice that the current trend is flat when it "should be" skyrocketing.

Given how much our culture is focused on the issue of CO_2-induced global warming, it is striking how little warming there has been.

But most striking to me are the data on how *dangerous* the climate has become over the last few decades, during a time when all

Figure 1.8: Global Warming Since 1850—the Full Story

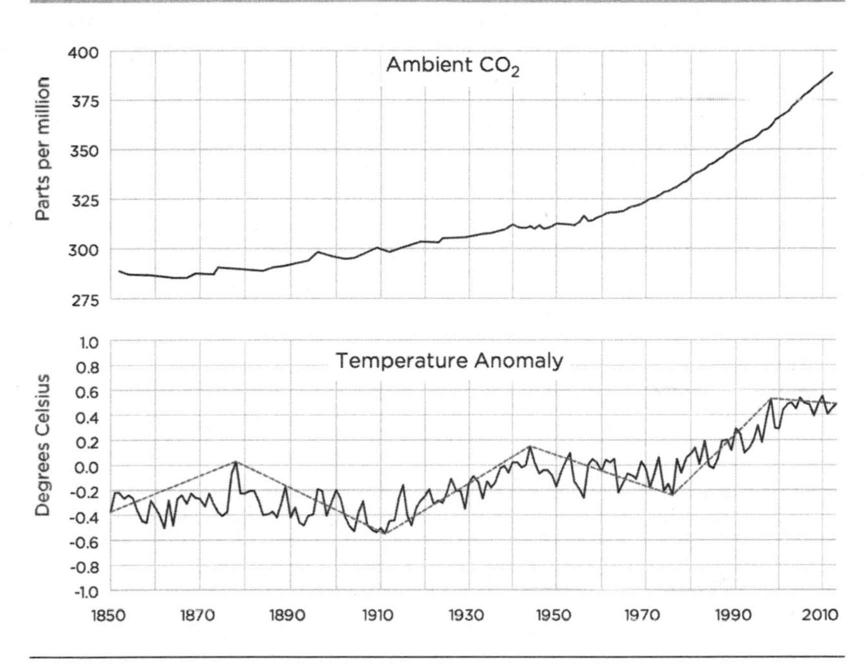

Sources: Met Office Hadley Centre HadCRUT4 dataset; Etheridge et al. (1998); Keeling et al. (2001); MacFarling Meure et al. (2006); Merged Ice-Core Record Data, Scripps Institution of Oceanography

of the predictions said that the Earth would become progressively more deadly. The key statistic here, one that is unfortunately almost never mentioned, is "climate-related deaths." I learned about this statistic from the work of the prolific global trends researcher Indur Goklany, who tracks changes over time in how many people die from a climate-related cause, including droughts, floods, storms, and extreme temperatures.[40]

Before you look at the data, ask yourself: Given what you hear in the news about the climate becoming more and more dangerous, what would you expect the change in the annual rate of climate-related deaths to be since CO_2 in the atmosphere started increasing significantly (about eighty years ago). When I speak at colleges, I sometimes get answers such as five times, even a hundred times greater death rates. And from the headlines, it does look as though the tragedies like Superstorm Sandy are the new normal.

The data say otherwise.

In the last eighty years, as CO_2 emissions have most rapidly escalated, the annual rate of climate-related deaths worldwide *fell* by an incredible rate of 98 percent.[41] That means the incidence of death from climate is *fifty times* lower than it was eighty years ago.

Figure 1.9: More Fossil Fuels, Fewer Climate-Related Deaths

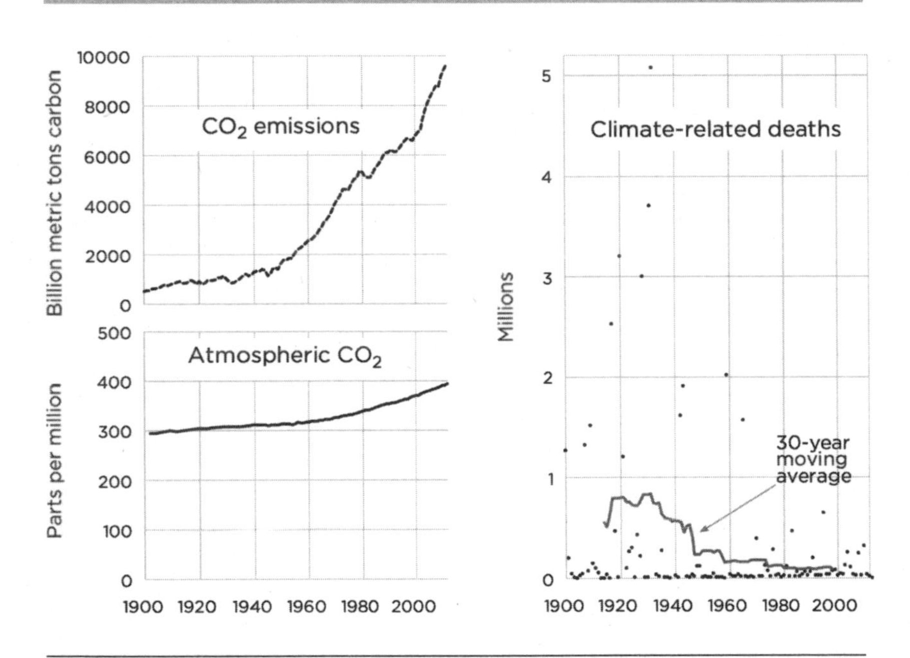

Sources: Boden, Marland, Andres (2013); Etheridge et al. (1998); Keeling et al. (2001); MacFarling Meure et al. (2006); Merged Ice-Core Record Data, Scripps Institution of Oceanography; EM-DAT International Disaster Database

The first time I read this statistic, I didn't think it was possible. But my colleagues and I at the Center for Industrial Progress have mined the data extensively, and it is that dramatic and positive. Because the numbers are so startling, in chapter 5 I'll explain them in depth.

Once again, the leading experts we were told to rely on were 100 percent wrong. It's not that they predicted disaster and got half a disaster—it's that they predicted disaster and got dramatic *improvement*. Clearly, something was wrong with their *thinking* and we need to understand what it is because they are once again telling us to stop using the most important energy source in our civilization. And we are listening.

Why did so many predict increasing climate danger when the reality turned out to be increasing climate safety as we used more fossil fuels? Once again, they didn't think big picture—they seemed to be looking only at potential risks of fossil fuels, not the benefits. Clearly, as the climate-related death data show, there were some major benefits—namely, the power of fossil-fueled machines to build a *durable* civilization that is highly resilient to extreme heat, extreme cold, floods, storms, and so on. Why weren't those mentioned in the discussion when we talked about storms like Sandy and Irene, even though anyone going through those storms was far more protected from them than he or she would have been a century ago?

WHAT'S AT STAKE

Imagine if we had followed the advice of some of our leading advisers then, many of whom are some of our leading advisers now, to severely restrict the energy source that billions of people used to lift themselves out of poverty in the last thirty years? We would have caused billions of premature deaths—deaths that were prevented by our increasing use of fossil fuels.

What happens if today's predictions and prescriptions are just as wrong? That would mean billions of premature deaths over the next thirty years and beyond. And the loss of a potentially amazing future.

Even if their predictions are partially right—certainly, fossil fuels have risks that we need to identify and quantify so as to minimize danger and pollution—we are in danger of making bad decisions because of the tendency to ignore benefits and exaggerate risks.

Today, proposals to restrict fossil fuels are more popular than ever. As mentioned earlier, the Intergovernmental Panel on Climate Change (IPCC) has demanded that the United States and other industrialized countries cut carbon emissions to 20 percent of 1990 levels by 2050—and the United States has joined hundreds of other countries in agreeing to this goal.[42] And the UN panel reassures us that "close to 80 percent of the world's energy supply could be met by renewables by mid-century if backed by the right enabling public policies . . ."[43] Around the world, it is fashionable to attack every new fossil fuel development and every new form of fossil fuel technology, from hydraulic fracturing ("fracking") in the United States to oil sands ("tar sands") in Canada.

To think about dire measures like this without seriously reflecting on the predictions and trends of the last forty years—and the thinking mistakes that led to those wrong predictions—is dangerous, just as it was dangerous for thought leaders to ignore the benefits of fossil fuels while focusing only on (and exaggerating) the risks. At the same time, we need expert guidance to know the *present-day* evidence about the benefits and risks of fossil fuels. History doesn't always repeat itself.

But how do we know what—and whom—to believe?

USE EXPERTS AS ADVISERS, NOT AUTHORITIES

Remember the question from my Greenpeace conversation: "So many experts predict that using fossil fuels is going to lead to catastrophe—why should I listen to you?" She—and we—shouldn't "listen" to anyone, in the sense of letting them tell us what to do.

To be sure, we absolutely need to consult experts. Experts are an indispensable source of information about the state of knowledge in specific fields—whether economics or energy or climate science—that we can use to make better decisions. But we can get this benefit only so long as the expert is clear about what he knows and how he knows it, *as well as what he doesn't know.*

Too often we are asked to take some action because an expert recommends it or because a group of experts favored it in a poll. This is a recipe for failure. We have already seen that the people revered as experts can be disastrously wrong, as Ehrlich was in his predictions from the seventies. Such errors are common, particularly among experts commenting on controversial political matters, where thinkers are rewarded for making extreme, definitive predictions. Think, for example, of all the economists who were convinced in 2007 and 2008 that the economy was healthy and who were advising people to take on more debt and purchase more property, inflating the real-estate bubble further and further, until it finally burst.

To avoid falling prey to this sort of "expert" advice, we need experts to explain to us how they reached their conclusions, and make sure they are not overstepping the bounds of their knowledge, which is incredibly common.

No scientist is an expert on everything; each specializes in some particular field. For example, a climate scientist might be a specialist in paleoclimatology (the study of using ancient evidence to deduce what ancient climates were like), and even then he might be an expert in only one period—say, the Cretaceous (one of the periods in which the dinosaurs lived). He is not going to be an expert in climate physics, and the climate physicist is also not an expert in human adaptation.

Whether our escalating use of fossil fuels is good or bad for us is a complex interdisciplinary question, and *everyone is a nonexpert in many relevant issues.* In this respect, we are all in the same boat. To reach an informed opinion, we need to draw on the work of experts in many fields, working to understand and evaluate their opinions and to interrelate them with one another and with our other knowledge.

Each of us is responsible for taking these steps—for doing his best to find the truth and to make the right decision. This means treating experts not as authority figures to be obeyed but as advisers to one's own independent thought process and decision making. An adviser is someone who knows more than you do about the specifics but knows only *part* of what you need and can be wrong. An honest and responsible expert recognizes this, and so he takes care to explain his views and his reasons for them clearly, he is upfront about any reasons there may be for doubting his conclusions, and he responds patiently to questions and criticism. He strives to give the public access to as much information as possible about his data, calculations, and reasoning. In this book, all the graphs are based on data collected from nonpartisan international sources (including arguably the three sources most respected by scholars: the World Bank, the International Energy Agency, and the BP Statistical Review of World Energy) and in-depth information about the graphs and how to re-create them can be found at www.moralcaseforfossilfuels.com.

SEEK THE BIG PICTURE

Ultimately, what we're after in examining the benefits and risks of fossil fuels is to know *big picture* how they affect human life and what to do going forward.

What experts in specific fields give us is knowledge that we can *integrate* into a big-picture assessment. For example, by learning from a combination of scientists and economists and energy experts, we can know how the risks of burning coal compare to the benefits of burning coal.

Looking at the big picture requires looking at *all* the benefits and risks to human life of doing something and of not doing it. To do otherwise is to be biased in a way that could be very dangerous to human life. One thing I noticed repeatedly when looking at the wrong predictions was a distinct bias against fossil fuels. The focus would be exclusively on the negatives of fossil fuels, which were often exaggerated, and not on their *positives*, which, given the results, were clearly overwhelming.

Often the cause of bias is an unacknowledged assumption.

For example, among those who disagree with catastrophic climate change predictions, it's a common assumption that it's *impossible* for man to have a catastrophic or even a significant impact on climate. For example, Indiana Congressman Todd Rokita says, "I think it's arrogant that we think as people that we can somehow change the climate of the whole earth . . ."—as if there is some preordained guarantee that we can't significantly affect the global climate system.[44] There isn't; whether we are or not can't be known without first examining the evidence.

On the other side of the issue, among those who agree with catastrophic climate predictions, it's a common assumption that there's something *inherently* wrong with man having an impact on climate. If you hold that assumption, you're likely to assume that the impact of man-made CO_2 emissions is very negative, even if the evidence showed it was actually mild or even positive.

We cannot assume things are good or bad. We must rigorously seek out the big-picture evidence—hence the last issue: being clear on exactly what we mean by good or bad.

NAME OUR STANDARD

Ultimately, when thinking about fossil fuels, we are trying to figure out the right thing to do, the right choices to make. But what exactly do we mean by right and wrong, good and bad? What is our *standard of value*? By what standard or measure are we saying something is good or bad, great or catastrophic, right or wrong, moral or immoral?

I hold human life as the standard of value, and you can see that in my earlier arguments: I think that our fossil fuel use so far has been a moral choice *because it has enabled billions of people to live longer and more fulfilling lives,* and I think that the cuts proposed by the environmentalists of the 1970s were wrong *because of all the death and suffering they would have inflicted on human beings.*

Not everyone holds human life as their standard of value, and people often argue that things are right or wrong for reasons other than the ways they benefit or harm human beings. For example, many religious people think that it is wrong to eat certain foods or to engage in certain sexual acts, not because there is any evidence that these foods or acts are unhealthy or otherwise harmful to human beings but simply because they believe God forbids them. Their standard of value is not human life but (what they take to be) God's will.

Religion is not the only source of nonhuman standards of value. Many leading environmental thinkers, including those who predict fossil fuel catastrophe, hold as their standard of value what they call "pristine" nature or wilderness—nature unaltered by man.

For example, in a *Los Angeles Times* review of *The End of Nature,* McKibben's influential book of twenty-five years ago predicting catastrophic climate change, David M. Graber, research biologist for the National Park Service, wrote this summary of McKibben's message:

> McKibben is a biocentrist, and so am I. We are not interested in the utility of a particular species or free-flowing river, or ecosystem, to mankind. They have intrinsic value, more value—to me—than another human body, or a billion of them. Human happiness, and certainly human fecundity, are not as important as a wild and healthy planet. I know social scientists who remind

> me that people are part of nature, but it isn't true. Somewhere along the line—at about a billion [*sic*] years ago, maybe half that—we quit the contract and became a cancer. We have become a plague upon ourselves and upon the Earth. It is cosmically unlikely that the developed world will choose to end its orgy of fossil-energy consumption, and the Third World its suicidal consumption of landscape. Until such time as *Homo sapiens* should decide to rejoin nature, some of us can only hope for the right virus to come along.[45]

In his book, McKibben wrote that our goal should be a "humbler world," one where we have less impact on our environment and "Human happiness would be of secondary importance."[46]

What is of primary importance? *Minimizing our impact on our environment.* McKibben explains: "Though not in our time, and not in the time of our children, or their children, if we now, *today*, limited our numbers and our desires and our ambitions, perhaps nature could someday resume its independent working."[47] This implies that there should be fewer people, with fewer desires, and fewer ambitions. This is the exact opposite of holding human life as one's standard of value. It is holding *human nonimpact* as one's standard of value, without regard for human life and happiness.

Earlier we saw that human beings are safer than ever from climate, despite whatever impact we have had from increasing the concentration of CO_2 in the atmosphere from .03 percent to .04 percent. And yet Bill McKibben and others call our present climate catastrophic. By what standard?

In his book *Eaarth*, McKibben argues that it's tragic for human beings to do anything that affects climate, even if it doesn't hurt human beings. He writes, referencing an earlier work:

> Merely knowing that we'd begun to alter the climate meant that the water flowing in that creek had a different, lesser meaning.

> "Instead of a world where rain had an independent and mysterious existence, the rain had become a subset of human activity," I wrote. "The rain bore a brand; it was a steer, not a deer."[48]

This means that something is morally diminished if human beings affect it.

If fossil fuels changed climate, but not in a way that harmed humans—or even helped them—would it be right to use them because of their benefits to human life?

On a human standard of value, the answer is absolutely yes. There is nothing intrinsically wrong with transforming our environment—to the contrary, that's our means of survival. But we do want to avoid transforming our environment in a way that harms us now or in the long term.

You might wonder how holding human life as your standard of value applies to preserving nature. It applies simply: preserve nature when doing so will benefit human life (such as a beautiful park to enjoy) and develop it when it will benefit human life. By contrast, if nonimpact, not human life, is the standard, the moral thing to do is always leave nature alone. For example, in the 1980s, India had an environmentalist movement, called the Chipko movement, that made it nearly impossible for Indians to cut down forests to engage in industrial development. It was so bad that a movement literally called Log the Forest emerged to counter it. As one Indian who tried to build a road said:

> Now they tell me that because of Chipko the road cannot be built [to her village], because everything has become *paryavaran* [environment]. . . . We cannot even get wood to build a house . . . *our haq-haqooq* [rights and concessions] *have been snatched away*. . . . I plan to contest the *panchayat* [village administrative body] elections and become the *pradhan* [village leader] next

> year. . . . My first fight will be for a road, *let the environmentalists do what they will.* [Italics in original][49]

This is the essence of the conflict: the humanist, which is the term I will use to describe someone on a human standard of value, treats the rest of nature as something to use for his benefit; the nonhumanist treats the rest of nature as something that must be served.

We always need to be clear about our standard of value so we know the goal we're aiming at. Aiming at human well-being, which includes transforming nature as much as necessary to meet human needs, is a lot different from aiming to *not* affect nature. The humanist believes that transforming nature is bad only if it fails to meet human needs; the nonhumanist believes that transforming nature is intrinsically bad and that doing so will inevitably somehow cause catastrophe for us in the long run.

Because many of the people predicting dire consequences from fossil fuel use avowedly do not hold a human standard of value and because the vast majority of discussions on the issue are not clear about the standard of value being used, we need to always ask, when we hear any evaluation: "By what standard of value?"

THE MORAL CASE FOR FOSSIL FUELS

In my experience, if we follow these principles to get a big-picture perspective on what will and won't benefit human life, the conclusion we'll reach is far more positive and optimistic than almost anyone would expect.

The reason is that the cheap, plentiful, reliable energy we get from fossil fuels and other forms of cheap, plentiful, reliable energy, combined with human ingenuity, gives us the ability to transform the world around us into a place that is far safer from any health hazards (man-made or natural), far safer from any climate change (man-made or natural), and far richer in resources now and in the future.

Fossil fuel technology transforms nature to improve human life on an epic scale. It is the only energy technology that can currently meet the energy needs of all 7+ billion people on this planet. While there are some truly exciting supplemental technologies that may rise to dominance in some distant decade, that does not diminish the greatness or immense value of fossil fuel technology.

Ultimately, the moral case for fossil fuels is not about fossil fuels; it's the moral case for using cheap, plentiful, reliable energy to amplify our abilities to make the world a better place—a better place *for human beings.*

That's where we will start. In chapters 2 and 3, I will make the case that no other energy technology besides fossil fuels can even come close to producing that energy for the foreseeable future (although several can be valuable supplements).

In chapters 4, 5, 6, and 7, I will make the case that just as energy dramatically improves our ability to deal with any aspect of life by using machines—increasing our mental capacities with computers, our medical capabilities with MRI machines, and our agricultural capabilities with high-powered farming equipment—so it dramatically improves our ability to make our environment healthier and safer from natural and man-made threats. The data clearly show that we have never had higher environmental quality and we have never been safer from climate, despite—no, because of—record fossil fuel use.

In chapter 8, I will make the case that fossil fuel use is not "unsustainable" but progressive—by using the best energy technology today and in the coming decades, we pave the way for fossil fuel technologies not only to harness the copious amounts of fossil fuels remaining in the ground, of which we have just scratched the surface, but also to create the resources and time necessary to develop the next great energy technology.

Finally, in chapter 9, I will make the case that we are at one of those points in history where we are at a fork between a dream and a nightmare and that the nightmare side is winning, thanks to decades of underappreciation of fossil fuels' benefits and massive misrepresentations of fossil fuels' risks. But the dream is absolutely possible. It just requires that we truly, to our core, understand the value of energy to human life.

2

THE ENERGY CHALLENGE: CHEAP, PLENTIFUL, RELIABLE ENERGY . . . FOR 7 BILLION PEOPLE

ENERGY AND LIFE

Tell me if this is motivating: This year humanity will use some 560 quadrillion BTU of energy, which averages out to around 215,000 BTU per person per day—and some people have access to less than 25,000 BTU per day.[1]

No?

Unfortunately, discussions of energy are often extremely abstract and technical, causing us not to think about energy in a very personal, meaningful way. Before I studied energy professionally, I thought of it mostly when I filled up my car, when I paid my power bill, and when I followed controversies about allegedly bad things the energy industry (usually the fossil fuel industry) was doing.

But the reality is that energy affects nearly every aspect of life. Almost nothing matters more to our lives, the lives of those you care about, and the lives of billions of others around the world than the existence of cheap, plentiful, reliable energy. To give you a sense of what I mean, here's a story from The Gambia about what electricity means to a woman having a child.

THE GAMBIA

June 2006

At 4 p.m. on a Saturday afternoon, I was startled when the lights came on; the lights never came on after 2 p.m. on the weekends. The adrenaline really kicked in when I was invited to observe an emergency cesarean section—a first for me. When the infant emerged I felt my heart racing from excitement and awe!

But no matter how many times the technician suctioned out the nose and mouth, the infant did not utter a sound. After twenty five minutes the technician and nurse both gave up. The surgeon later explained that the baby had suffocated in utero. If only they had had enough power to use the ultrasound machine for each pregnancy, he would have detected the problem earlier and been able to plan the C-section. Without early detection, the C-section became an emergency, moreover, the surgery had to wait for the generator to be powered on. The loss of precious minutes meant the loss of a precious life. At that time, in that place, all I could do was cry.

And later, when the maternity ward was too hushed, I cried again. A full-term infant was born weighing only 3.5 pounds. In the U.S., the solution would have been obvious and effective: incubation. But without reliable electricity, the hospital did not even contemplate owning an incubator. This seemingly simple solution was not available to this newborn girl, and she perished needlessly.

Reliable electricity is at the forefront of every staff members' thoughts. With it, they can conduct tests with electrically pow-

> ered medical equipment, use vaccines and antibiotics requiring refrigeration, and plan surgeries to meet patients' needs. Without it, they will continue to give their patients the best care available, but in a country with an average life expectancy of only 54 years of age, it's a hard fight to win.[2]

This story should remind us of how "unnatural" our lives are—and why that's a good thing. It's easy to take for granted that we have the ability to detect early problems with babies—not thinking that absent the machine that can detect them and the energy to power that machine, human beings past and present have lost untold millions of babies. It's easy to take for granted that we have the ability to keep a three-and-a-half-pound baby alive—not thinking that absent the machine that can incubate it and the energy to power that machine, most of people's beloved children who were underweight babies would have died.

This is a microcosm of the central idea of this book—that more energy means more *ability* to improve our lives; less energy means less ability—more *helplessness*, more suffering, and more death. Of course, this book is focused on *fossil fuel* energy—but only, as you'll see, because I believe that it is the most essential technology for producing energy for 7 billion people to improve their lives, at least over the next several decades. If there was a better form of energy and it was under attack in a way that wildly exaggerated its negatives and undervalued its positives, I'd be writing the moral case for *that* form of energy.

There are two facts about energy that are missing from our discussion: one, people around the world need much, much more energy, and two, it's extremely difficult to produce that energy cheaply and reliably.

MACHINE CALORIES

Humanity needs as much energy as it can get.

First: What exactly is energy? The technical definition is "the capacity to do work" but my favorite way to sum it up is with two words: "machine calories."

Every human being runs off the calories he or she consumes; those calories are our *energy*, our *ability to act*. If we run out of calories, we can't act—we die.

The same is true of the machines we use to improve our lives. Whether we're talking about the ultrasound and incubation machines that enable us to save babies, the computers that enable us to gain or discover knowledge, the planes that enable us to visit family members across the globe, or the factories that make it possible for all of those things to be affordable, every aspect of our lives is improved dramatically by machines. Those machines live on energy—*their* ability to act—and without energy, they are the same as the energy-starved machines that can't save the Gambian babies: useless.

And we desperately need machines to do work for us because we are naturally *very weak*. Without machines to help us, we don't have anywhere near all the energy that we need to survive and flourish.

The average human being needs about 2,000 calories a day to give him enough energy to do everything he needs to do—from going to the office to taking a walk to manual labor to sleeping. That's equal to 2,326 watt-hours, which is the amount of energy it takes to power a 100-watt lightbulb for 23.26 hours. Essentially, your body uses the same amount of energy as a 100-watt lightbulb. Pretty interesting, right?

The more physical work you need to do, the more calories you use. A farmer doing vigorous physical work for a day might exert 4,000 calories.[3] An Olympic athlete like Michael Phelps might use 8,000 calories of energy a day.[4]

The more energy you are using at any time, the more *power* you are exerting. Power is defined as the rate of energy use. Power is energy in action; the gasoline is the energy, the engine turns it into power.

And here's where the problem of human weakness comes in. We are not very powerful—about one tenth as powerful as a horse that's one two-hundredth the power of the average car—and thus we can use only so much energy and do only so much work, not nearly enough for a good standard of living.

The story of energy for over 99 percent of history is that human beings couldn't get enough of it to live, and if they could, they could make very limited use of it, because they lacked *power*. Thus they spent their lives engaging in grueling physical labor just to keep their bodies going long enough to engage in the next day of grueling physical labor.

Now if we were all like Superman, it would be a different story. Imagine if Superman, instead of devoting himself to saving Lois Lane and others, decided to help poor countries industrialize. He would be amazing! Superman's superpower, after all, is *power*. He is a high-powered machine that stores a lot of energy in his body. He can melt iron, forge steel, plow fields, build buildings, even run an electrical system by turning some sort of especially large crank. *He could transform any place for the better.*

And so can we, with enough energy and high-powered machines. Using human ingenuity, we have made ourselves into supermen.

Consider the amount of energy at the average American's disposal. The average American's total machine energy use is 186,000 calories a day—ninety-three humans (or twenty-three Michael Phelpses)![5] This is one of the greatest achievements in human history. In the past, before modern energy technology, the main way to overcome the problem of human weakness was putting others into a state of servitude or slavery—which meant that only some could prosper, and at the great expense of others. But with machine energy and machine servants, no one has to suffer; in fact, the more people, the merrier.

The most memorable summary I've read about this amazing development is by economist Milton Friedman:

> Industrial progress, mechanical improvement, all of the great wonders of the modern era have meant little to the wealthy. The rich in ancient Greece would have benefited hardly at all from modern plumbing—running servants replaced running water. Television and radio—the patricians of Rome could enjoy the leading musicians and actors in their home, could have the leading artists as domestic retainers. Ready-to-wear clothing, supermarkets—all these and many other modern developments would have added little to their life. They would have welcomed the improvements in transportation and in medicine, but for the rest, the great achievements of western capitalism have redounded primarily to the benefit of the ordinary person. These achievements have made available to the masses conveniences and amenities that were previously the exclusive prerogative of the rich and powerful.[6]

"Running servants replaced running water"—I'll never forget that.

THE ENERGY CHALLENGE: CHEAP, PLENTIFUL, RELIABLE, SCALABLE

If our ability to act to improve our lives depends on energy, we have an epic challenge.

There are 7 billion people in the world, but 1.3 billion have no electricity.[7] Over 3 billion are classified as not having "adequate electricity"—a threshold that is far less than we enjoy and take for granted.[8] For everyone to have as much access to energy as the average American, the world's energy production would have to quadruple.[9] And we Americans would benefit greatly from even more cheap, plentiful, reliable energy.

So where are we going to get it from?

In this chapter and the next, we're going to examine every major energy technology, including all the non–fossil fuel sources of energy, to get an idea of how much they can contribute to energy production going forward. This is important because, assuming you can do it safely, the more energy production, the better and also because there are concerns about the future supply of fossil fuels and the present and future *risks* of fossil fuels. We'll cover future supply in the next chapter and in chapter 8, and the risks in chapters 4–6, but as a matter of principle, anytime we are worried about the risks of one way of doing things (here, using fossil fuel energy) we need to know the benefits and risks of the alternatives.

Nineteenth-century coal technology is justifiably illegal today. The hazardous smoke that would be generated is now *preventable* by far more advanced, cleaner coal-burning technologies. But in the 1800s, it was and should have been perfectly legal to burn coal this way—because the alternative was death by cold or starvation or wretched poverty.

By the same token, the degree of risk we would theoretically be willing to accept from fossil fuels will depend in large part on *what the alternatives are*. Let's say—and I am completely making this up—we could prove that burning fossil fuels will cause ten times more hurricanes for the next fifty years. Should the government take action? Well, if there is a technology that is more affordable and can scale to produce cheaper, more plentiful, reliable energy for 7 billion people, then quite possibly. But if there is no equal or superior alternative, then any government action against fossil fuels, let alone the 50–95 percent bans over the next several decades that have been proposed, is a *guaranteed early death sentence* for billions—

we would be willing to accept ten times more hurricanes if we had to. Energy is that important.

To get a sense of where things stand today and where they stood in the past, when "renewables" were predicted to be the future, let's look at how much energy use comes from what sources.

Figure 2.1: The Truth About Global Energy Use

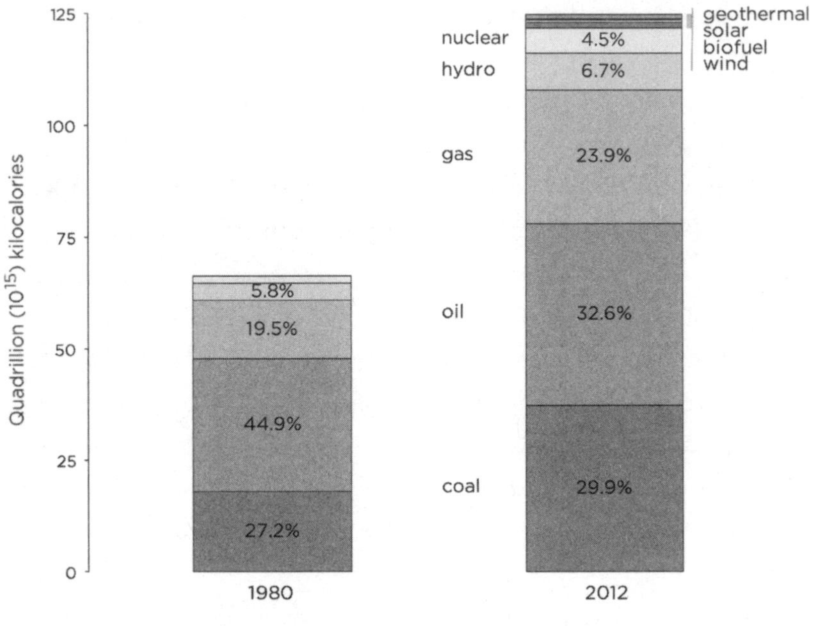

Quadrillion (10^{15}) kilocalories
125
100
75
50
25
0
1980
5.8%
19.5%
44.9%
27.2%
2012
geothermal
solar
biofuel
wind
nuclear 4.5%
hydro 6.7%
gas 23.9%
oil 32.6%
coal 29.9%

Source: BP, Statistical Review of World Energy 2013, Historical data workbook

Note the difference. Solar and wind produce a combined 1 percent of the energy we use, whereas fossil fuel energy—coal, oil, and natural gas—produces 86 percent, more than five times all other sources combined. That 86 percent is only 7 percent less than 1980's 93 percent. But the total is what matters most—note that our total fossil fuel use is now far, far greater. *Other sources of energy, particularly nuclear and hydro, have been supplements, not replacements for fossil fuels.*

And note that in many ways people have been discouraged from using fossil fuels. For the last thirty years, governments around the world—particularly European governments like Germany, Spain, and Denmark—have gone out of their way to promote non-fossil forms of energy, such as solar, wind, and biofuels. Nevertheless, fossil fuels have remained the energy source of choice.

Why? Or to put it in reverse, why is so much energy *not* made from alternatives?

THE HAZELNUT ENERGY PROBLEM

The simple answer is: because it's a really, really, really hard challenge to produce cheap, plentiful, reliable energy for billions of people—and the fossil fuel industry is the only one, by a mile, that's figured out a solution. (Although there's one source of energy that may well outcompete fossil fuels in three to five decades—stay tuned.)

A brilliant illustration of this appeared on, of all places, *Saturday Night Live* a few years ago. The host of the "Weekend Update" segment at the time, Jimmy Fallon, commented on a plan to use oil derived from hazelnuts to power a car. I have no doubt that this could work technically—vegetable oil and petroleum oil are extremely similar chemically. But I wasn't excited, and neither was Fallon:

> *New Scientist* magazine reported that in the future, cars could be powered by hazelnuts. That's encouraging, considering an eight-ounce jar of hazelnuts costs about nine dollars. Yeah, I've got an idea for a car that runs on bald eagle heads and Fabergé eggs.[10]

I thought that was brilliant. But here's the question I wished Fallon, a member of Artists Against Fracking and thus a public opponent of fossil fuels, had asked: Why are "renewable" hazelnuts so expensive? After all, their energy comes from the sun, which is free, right?

He probably would have responded that while the sun is free, there were other factors in the *process* of producing hazelnuts that make them expensive.

And there are.

Here's a key principle for understanding what makes energy, or anything else, cheap and plentiful. *For something to be cheap and plentiful, every part of the process to produce it, including every input that goes into it, must be cheap and plentiful.* With hazelnuts, not only do you have, as in any process, materials, machines, and manpower, you have a huge limiting factor in that the *land* needed is far from plentiful. Hazelnuts require land with a unique combination of rainfall or irrigation, mild summer climate and cold winter climate, and fertile soil. This happens overwhelmingly in one place, Turkey, which dominates the market, and this ideal hazelnut habitat generates only one crop a year.[11]

What we can call the hazelnut problem comes up over and over again with most of the alternatives to fossil fuels. In some cases, they may be cheap and reliable in small quantities—some people use French fry oil to power their cars—but making them cheap and reliable in large quantities, quantities sufficient to power the lives of billions of people, is a major feat.

Just as it's a mistake to assume that because the sun is free, solar-powered hazelnuts will be cheap, so it is a mistake to assume that solar-powered energy can or will be cheap. Whether that's true or not depends on all the materials, manpower, and machines involved in the entire *process* of harnessing the sun's power.

Every energy process requires taking a form of raw energy—there is no ready-made machine energy—and *transforming* it into usable form so that it becomes the heat in our homes, the mechanical power of our cars, and the electricity that powers the Internet. This is a process that takes time and resources, and the key is to make it take as little time and as few resources as possible, so that it can be *workable* (including reliable), cheap, and plentiful.

Workable is a challenge. Cheap and plentiful are an incredible challenge.

Hazelnut energy is workable; it just isn't likely going to be cheap and plentiful.

Another related challenge is dealing with risks and by-products. Every time energy is transformed, there is the risk of something going wrong (explosion, electrocution), and there are by-products that can be harmful (such as sulfur dioxide from coal or radioactive waste generated when mining the metals that go into windmills).

Let's look at which technologies work worst and best at providing cheap, plentiful, reliable energy—and by plentiful I mean on the scale of *billions* of people. For each one, I'll give a brief summary of how it works, how successful it has been at producing cheap, plentiful, reliable energy, and how it is positioned for the future. I'll start with the most culturally popular energy technologies: solar and wind.

THE EFFICIENCY PROBLEMS OF SOLAR AND WIND: DILUTENESS AND INTERMITTENCY

Solar and wind energy both work with energy flowing directly from the sun; solar through sunlight and wind through the sun's heating of different parts of the atmosphere, which is the main cause of wind.

Solar energy typically works in one of two ways: solar photovoltaic (abbreviated solar PV) and concentrated solar power (CSP). Solar PV generates electricity through a phenomenon, discovered around 1839 by Edmond Becquerel, called the photovoltaic effect, by which certain materials emit electrons when hit by light. Through extremely impressive feats of engineering involving precision (and often expensive) materials, solar PV can generate an electric current when it is exposed to sunlight. The first "solar cell" was patented in the United States in *1888*. CSP, by contrast, concentrates sunlight into heat, much as a child with a magnifying glass does when he uses the sun to ignite a dried leaf. Using, in effect, a massive array of magnifying glasses (in this case, mirrors), CSP concentrates sunlight into heat, which is used to heat a liquid, which generates steam that can power an engine.

Wind electricity works when high-velocity wind turns the blades of a wind turbine, which are connected to a generator that converts the wind's power into electric current.

In practice, solar and wind technologies have, as we saw before, produced very, very little energy.

The top five countries ranked by solar energy consumption are Germany, Italy, Spain, Japan, and China. The percentage of each country's electricity that comes from solar energy is, respectively: 4.5 percent, 6.3 percent, 4.0 percent, .09 percent, and .6 percent.[12]

The top five countries ranked by wind consumption are the United States, China, Spain, Germany, and India. Faring slightly better than solar, the percentage of each country's electricity that comes from wind energy is, respectively: 3.3 percent, 2.03 percent, 16.5 percent, 7.44 percent, and 2.96 percent.[13] (If this seems impossibly low, because we frequently hear numbers such as "50 percent solar and wind," stay tuned.)

Don't let the 16.5 percent in Spain mislead you. Spain suffered financial devastation from its investment in wind, among other bad investments.[14] But more important, certain fundamental problems with solar and wind mean that the more energy they attempt to produce, the more of a problem they create.

Why?

The basic problem is that the *process* for solar and wind to generate reliable electricity requires so many resources that it has never been cheap and plentiful. In fact, *modern solar and wind technology do not produce reliable energy, period.*

Traditionally in discussions of solar and wind there are two problems cited: the diluteness problem and the intermittency problem.

The diluteness problem is that the sun and the wind don't deliver concentrated energy, which means you need a lot of materials per unit of energy produced. For solar, such materials can include highly purified silicon, phosphorus, boron, and compounds like titanium dioxide, cadmium telluride, and copper indium gallium selenide.[15] For wind, they can include high-performance compounds (like those used in the aircraft industry) for turbine blades and the rare-earth metal neodymium for lightweight, high-performance magnets, as well as the steel and concrete necessary to build thousands or tens of thousands of structures as tall as skyscrapers.[16]

Figure 2.2 indicates how steel (and iron) intensive it is to generate electricity from wind as compared with coal, nuclear, or natural gas.

Figure 2.2: Steel and Iron Required per Megawatt for Wind, Coal, and Natural Gas

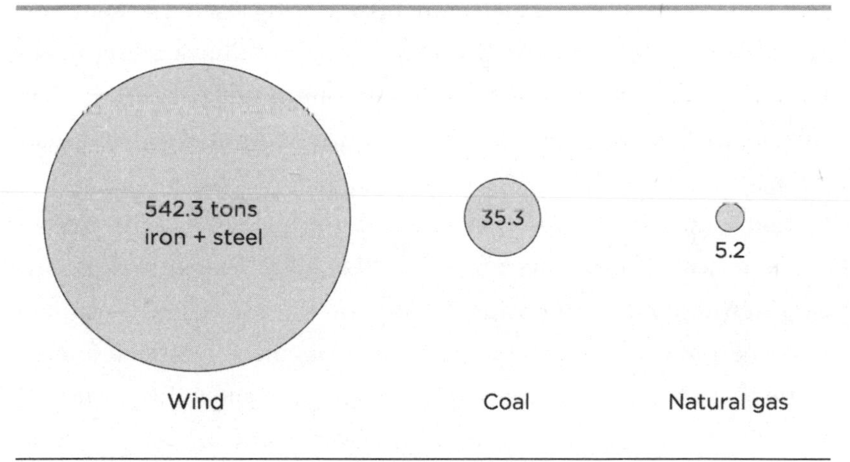

Sources: ALPINE Bau GmbH, July 2014; Peterson, Zhao, Petroski (2005); Wilburn 2011

Such resource requirements are a big cost problem, to be sure, and would be one even if the sun shone all the time and the wind blew all the time. But it's an even bigger problem that the sun and wind don't work that way. That's the real problem—the intermittency problem, or more colloquially, the unreliability problem.

As we saw in the Gambian hospital, it is of life and death importance that energy be reliable. There are some situations where it isn't, to be sure, and solar has a place there—such as solar hot water heaters or swimming pool heating systems. But for just about everything we do, reliable, on-demand energy is vital—and without it, our electricity grid blacks out.

We know from experience that the sun doesn't shine all the time, let alone with the same intensity all the time, and the wind doesn't blow all the time—and leaving aside the assurance that the sun will be "off" at night, they can be extremely unpredictable.

To hear opponents of fossil fuels discuss the issue, though, the unreliability of solar and wind is no obstacle at all, as evidenced by, above all, the success of Germany in powering itself via solar and wind. In late 2012, Bill McKibben described "what's going on in Germany" as "un-[expletive]-believable" and said "there were days this month [December] when they got half their energy from solar panels."[17]

And it appears that the news is just getting better. The Center for American Progress reported on May 13, 2014, that "Germany Sets New Record, Generating 74 Percent of Power Needs from Renewable Energy."[18] But taking a look at Germany's official energy statistics tells a very different story. Figure 2.3 shows how much of Germany's energy actually came from solar and wind throughout 2013, compared with how much was typically needed during each month.[19] Notice how unreliable the quantity of solar and wind electricity is. Wind is constantly varying, sometimes disappearing nearly completely, and solar produces very little in the winter months, when Germany most needs energy.

Figure 2.3: Solar and Wind Provide a Small, Unreliable Fraction of Germany's Electricity

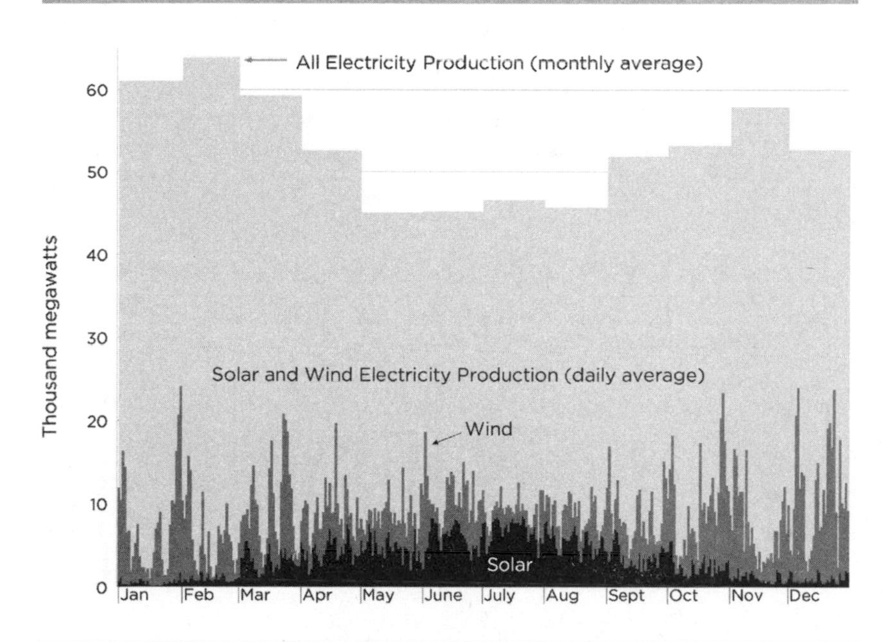

Sources: European Energy Exchange Transparency Platform Data (2013); Federal Statistical Office of Germany

How, then, can so many say that solar or wind generates over 50 percent of Germany's energy? What they are referring to is the fact that because solar and wind are so variable, at any given *moment* solar can generate 50 percent of the electricity being used. It can also generate 0 percent of the electricity generated at any given moment. Here is a graph of German solar and wind production in April 2013 based on the average amount of electric power generated every fifteen minutes. Notice that sometimes the combined output of solar and wind is relatively high—and sometimes it is nothing; that is the nature of an intermittent, unreliable source.

Figure 2.4: Solar and Wind: The Closer You Look, the Less Reliable They Are

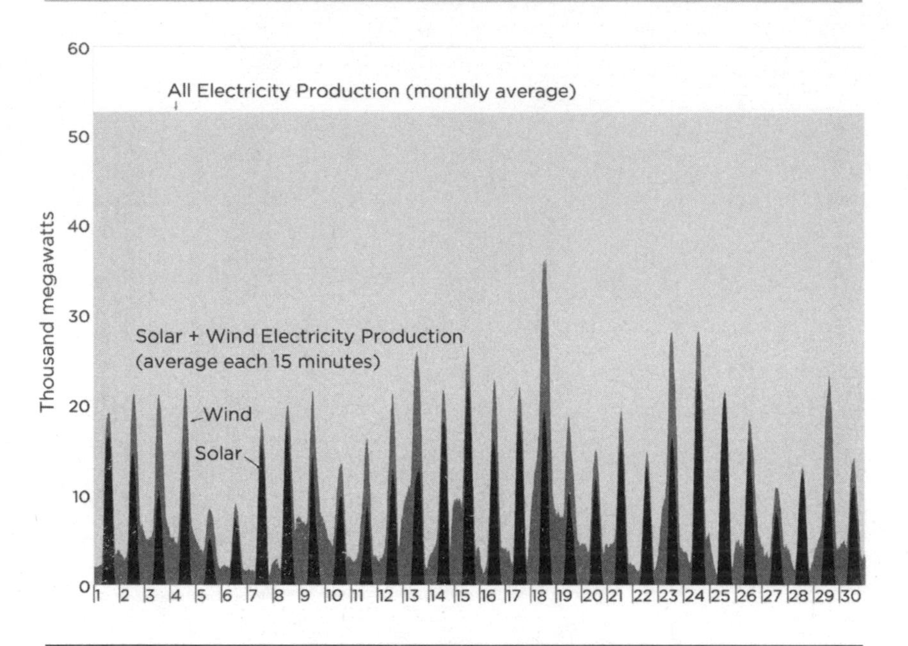

Sources: European Energy Exchange Transparency Platform Data (2013); Federal Statistical Office of Germany

As you look at the jagged and woefully insufficient bursts of electricity from solar and wind, remember this: *some reliable source of energy needed to do the heavy lifting.* In the case of Germany, much of that energy is coal. As Germany has paid tens of billions of dollars to subsidize solar panels and windmills, fossil fuel capacity, especially coal, has not been shut down—it has increased.[20]

Why? Because Germans need more energy, and they cannot *rely* on the renewables.

In a given week in Germany, the world leader in solar and number three in wind, their solar panels and windmills may generate less than 5 percent of needed electricity.[21] What happens then? Reliable sources of energy, in Germany's case coal, have to produce more electricity. For various technical reasons, this is even more inefficient than it sounds. For example, because the reliable sources have to move up and down quickly to adjust to the whims of the sunlight and wind, they become inefficient—just like your car in stop-and-go-traffic—which means more energy use and incidentally more emissions (including CO_2). And what about when there's a particularly large amount of sunlight or wind? For an electric grid, too much electricity will cause a blackout just as too little will—so then Germany has to shut down its coal plants and be ready to start them up again (more stop and go). In practice they often have so much excess that they have to *pay* other countries to take their electricity—which requires the other countries to inefficiently decelerate *their* reliable power plants to accommodate the influx. This is obviously not scalable; if everyone's electrical generation was as unreliable as Germany's, there would be no one to absorb their peaks.

The only way for solar and wind to be truly useful, reliable sources of energy would be to combine them with some form of extremely inexpensive mass-storage system. No such mass-storage system exists, because storing energy in a compact space itself takes a lot of resources. Which is why, in the entire world, there is not one real or proposed independent, freestanding solar or wind power plant. All of them require backup—except that "backup" implies that solar and wind work most of the time. It's more accurate to say that solar and wind are parasites that require a host.

Here's an analogy. Imagine you have a company of highly productive, efficient, reliable workers. Then there is an initiative to bring in "renewable" workers, who will supposedly live forever, but they are expensive and you don't know when they'll show up. A document produced by them is not as valuable as a document produced by someone else—because you don't know when theirs will arrive. A company can handle a few such workers, but it can't be run by them.

I remember watching an interview of a doctor in Kenya who had to try to run his practice with renewable energy. His clinic was run on solar and could not produce enough electricity to keep both the lights and the refrigerator on at all times, so he had to choose one or the other. When he tried switching on both, an alarm sounded, signifying "out of power."[22] Out of power is exactly the danger to the extent we try to substitute solar and wind for fossil fuels.

Another Kenyan, James Shikwati of the Inter Region Economic Network, explains why he resents programs to encourage underdeveloped countries to use solar or wind.

> The rich countries can afford to engage in some luxurious experimentation with other forms of energy, but for us we are still at the stage of survival. I don't see how a solar panel is going to power a steel industry, how a solar panel is going to power a railway network, it might work, maybe, to power a small transistor radio.[23]

Why do environmentalists focus so much on solar and wind, despite their intractable problems? The traditional explanation is that they don't generate CO_2—leaving aside the coal and oil needed to manufacture them (you can't build a windmill with a windmill). But as we'll see later, there are other, much more scalable forms of energy that don't generate CO_2 (hydroelectric and nuclear), which environmental leaders *oppose.*

Regardless of one's views on the risks of fossil fuels, it is profoundly irresponsible to claim, as many advocates of solar and wind do, that they are powering Germany, let alone supplying 50 percent of the power. Energy is a life and death issue—it is not one where we can afford to be sloppy in our thinking and seize upon statistics that seem to confirm our worldview.

It seems that there's more focus on getting energy directly from the sun, which is often considered "natural," than there is on getting it in a way that will maximize human life. It is deeply irresponsible and disturbing that environmental leaders are telling us to deprive ourselves of fossil fuels on the promise of what can charitably be described as a highly speculative experiment, and can less charitably be described as an ill-conceived, resource-wasting, perennial failure.

There is one much more reliable source of renewable energy that is endorsed by many environmental leaders, though with some reluctance: biomass energy. For example, in order to meet renewability mandates, which usually exclude hydroelectric power, Germany and various other countries are turning to a renewable biomass fuel from the past to make up for the fact that solar and wind scale so poorly: wood.

THE PROBLEMS WITH BIOMASS: PROCESSING AND SCALABILITY

Biomass energy is derived from plant or animal matter, whether wood, crops, crop waste, grass, or even manure. Biomass includes biofuels, which are liquid fuels, usually alcohol, derived from these sources and used for mobile power. Other forms of biomass are used for fixed electrical power or directly for heat (such as wood or animal dung burned to stay warm).

In practice, biomass has, like solar and wind, produced a small amount of energy worldwide—although considerably more than solar and wind.

Why?

Biomass is renewable and natural, because the energy comes from the sun—but not all the inputs in the process can scale. It resembles hazelnut energy; in fact, hazelnut energy is a form of biomass energy.

To its credit, biomass has a storage system, unlike solar and wind—plants store energy from the sun through photosynthesis. The problem is, it takes a lot of resources to grow them—namely the resources involved in farming, including large amounts of energy, land, machinery, water, fertilizer—just like it takes a lot of water to build solar and wind installations. But while solar and wind installations can be built in many places (though part of their problem is that northern and southern latitudes don't give them good sunlight for many hours), biofuels need to be grown on relatively scarce farmland, which starts to bring us into hazelnut energy territory. It means that biomass scales badly—often, the more of it we try to produce, the more scarce and expensive the inputs become, and the more expensive our energy becomes.

Biofuels like ethanol from corn or sugarcane, or biomass from wood, compete with cropland or forest land, driving prices up for both fuel and food.[24] Scalability has been the problem for every biofuel that works (the Bush administration tried to force us to use cellulosic ethanol, a form of ethanol from nonfood sources that has been promoted since the 1920s but still doesn't work) at a smaller scale. But even if nonfood biomass worked better than it does, it would still be extremely resource intensive to regrow over and over.

A thought: Throughout history it has been a challenge for human beings to produce enough crops to feed us, because agriculture requires a lot of resources just to produce our meager number of calories. We need many dozens of times as many calories for our machines as we do from our food! If we could eat oil or electricity, we would, because it's much cheaper per unit of energy. Why should we feed human food to machines with hundreds of times our appetites?

Already, the increased use of biomass energy has strongly correlated with a rise in food prices—see Figure 2.5. The idea of scaling it ten times or more, to even make a dent in fossil fuels' energy production, is unthinkable, given all of the evidence we have.

According to a recent report from the United Nations, *The State of Food Insecurity in the World,*

High and volatile food prices are likely to continue. Demand from consumers in rapidly growing economies will increase, population continues to grow, and any further growth in biofuels will place additional demands on the food system.[25]

Figure 2.5: Comparison of Food Price Index to Biofuel Production

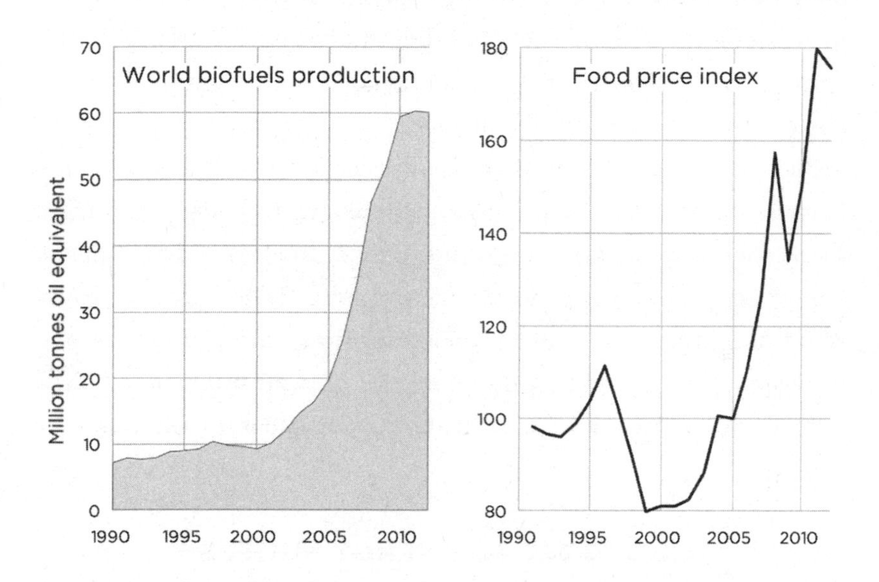

Sources: index mundi Commodity Food Price Index, 2014; BP, Statistical Review of World Energy 2013, Historical data workbook

Biomass energy is not providing scalable energy, but it is making it difficult for farmers to provide scalable *food.*

Here's the bottom line with solar, wind, and biofuels—the three types of energy typically promoted in renewables mandates. There is zero evidence that solar, wind, and biomass energy can meaningfully *supplement* fossil fuel energy, let alone replace it, let alone provide the energy *growth* that is desperately needed. If, in the future, those industries are able to overcome the many intractable problems involved in making dilute, unreliable energy into cheap, plentiful, reliable energy on a world scale, that would be fantastic. But it is dishonest to pretend that anything like that has happened or that there is a reason to think it will happen.

To be sure, solar, wind, and biomass may have their utility for niche uses of energy. If you're living off the grid and can afford it, an installation with a battery that can power a few appliances might be better than the alternative (no energy, or frequently returning to civilization for diesel fuel), but they are essentially useless in providing cheap, plentiful, reliable energy for 7 billion people—and to try to rely on them would be deadly.

And yet our leaders propose massive bans on fossil fuels with the promise that *these* radically inferior technologies will be replacements. That reflects an ignorance of, or indifference to, the need for efficient energy and the value of cheap, plentiful, reliable energy. Any leader who is thinking about making policy decisions with our energy, and ultimately the energy and therefore the opportunities of 7 billion people, had better take the truth about renewables into account.

THE SECRET TO ENERGY SUCCESS: NATURALLY CONCENTRATED, STORED, PLENTIFUL ENERGY

One lesson of the failure of renewables is that *renewable* is not a useful criterion for a good energy source. It says only that one of the inputs is derived from the sun; it says nothing about how long the other inputs will last, and, most important, it says nothing about whether the technology can generate energy that is cheap, plentiful, and reliable. There's no reason to aspire to use an energy technology that we will use forever. The real question is: For the relevant time horizon, what's the most efficient combination of elements that we can transform efficiently into the kind of energy we need in a way that is cheap, plentiful, and reliable?

And so far in history, there has been one necessary ingredient to that: instead of spending huge amounts of resources concentrating and storing a dilute and intermittent source, working with a source that nature has already concentrated and stored for us—such as water (hydropower), the forces holding an atom together (nuclear power), or the powerful chemical bonds of the copious amounts of ancient, dead plants lying around from previous eons (fossil fuels).

It is their preconcentrated, prestored, plentiful energy content that has made fossil fuels—and to a much less but still important extent hydroelectric power and nuclear power—cheap, plentiful, reliable energy sources.

HYDRO TECHNOLOGY: CHEAP, RELIABLE, MEDIUM-SCALE ENERGY

If you've ever been in a rapidly flowing river, you can *feel* the energy stored in the moving water. Hydroelectric energy technology transforms some of the power of that flowing water into usable, cheap, reliable electricity using a turbine—much like a wind turbine, except driven by a far more powerful and reliable force. Often a dam is used to store water near the source of a river and precisely control the downward flow.

Historically, hydropower has faced two types of limitations that have prevented it from producing much more than 6 percent of the world's power.[26] One category is natural limitations; the other is political limitations.

The main limitation of hydroelectric power is there aren't nearly enough suitable water sites for it to be a global source of energy. In China and Brazil, the top two consumers of hydropower, you can get a lot of electricity from it; in Nebraska, you can't.[27] The United States has maintained a fairly constant hydropower consumption because we've run out of rivers to dam (which is unfortunate, because hydropower lasts for decades; the Hoover Dam was built in the 1930s).

But there is considerably more opportunity to develop hydro around the world. Based on the number of dammable rivers left, the International Energy Agency estimates that hydroelectricity has the technical potential to grow by 92 percent in Africa and 80 percent in Asia.[28] Worldwide, according to an estimate by the International Energy Agency, hydro has the technical potential to produce twice as much energy as it does today; it is currently around 6 percent of global production.[29] That is an exciting prospect . . . but not for most prominent environmental groups, whom you might think would welcome a four times greater supply of cheap, reliable, non-CO_2-emitting hydroelectric energy.

Environmental activists have spent decades shutting down as many hydroelectric dams as possible, particularly large hydroelectric dams, despite hydro's proven track record as a cheap, reliable source of CO_2-free power, in the name of protecting species of fish, free-flowing rivers, and other justifications that focus on nonhuman nature.[30] The Sierra Club on its list of accomplishments on its Web site lists dams it has prevented or shut down.[31]

If the standard is improving human life, those who believe that catastrophic climate change is coming unless we reduce CO_2 emissions should favor damming every possible river to generate reliable CO_2-free power. And for those who don't believe CO_2's climate impact is a major problem, there's still a huge burden of proof on anyone to justify depriving people of a cheap, plentiful, reliable source of energy.

NUCLEAR TECHNOLOGY: RELIABLE, SCALABLE . . . CHEAP?

With hydroelectric, we saw that a *naturally concentrated, stored source* of energy was a big benefit. This is the reason why the potential of nuclear power has enchanted many in the energy world, this author included.

If natural concentration is a benefit, there is no more naturally concentrated energy source than the uranium or other radioactive metals used to generate nuclear power. Oil is an amazingly concentrated source of energy, which is why it is the transportation fuel of choice. Well, the concentration of energy in uranium is more than a *million* times that of oil and 2 million times that of coal—although given current technology, in practice it "only" delivers *thousands* of times more energy per unit of input.[32]

Nuclear's presence in generating energy around the world is slowly growing. There are two factors, which can be hard to separate, that hold back nuclear's progress: the difficulty of doing it cheaply and the perceived difficulty of doing it safely.

While many feel that the focus in the nuclear process should be on safety, I think the evidence shows that the real controversy should be on *price*.

Recall that to produce cheap, plentiful, and reliable energy, every element of the energy production process has to be cheap, plentiful, and reliable. Nuclear power uses uranium, which exists in enormous quantities around the world, and can also use thorium, an even more abundant material. Even using current technology, we are talking about time horizons upwards of thousands of years. The trickier part of the process is transforming that material into energy, which is much more complex than, say, burning natural gas to generate electricity. It involves producing energy by releasing the immense forces within a radioactive atom. Absent proper safety technology, human exposure to large amounts of radioactivity can lead to radiation poisoning or, in the longer term, cancers. At the same time, below a certain threshold, radioactivity is not harmful; we ourselves are radioactive and emit radioactivity. Unfortunately, radioactivity is commonly viewed as deadly *as such,* so critics of nuclear power can cite amounts of radiation coming from, for example, the Fukushima accident, and it sounds scary—even though the amount is not enough for anyone to die now (of radiation poisoning) or in the future (from cancer).

The issue of nuclear safety is full of so much rhetoric and emotion that it can be hard to sort through. But as a starting point, let's ask: How do we know how safe it is? I think the most reliable indication of a technology's safety is how many deaths it has caused per unit of energy produced. In the free world, nuclear power in its entire commercial history has not led to a single death—including from much-publicized failures at Three Mile Island and Fukushima.[33]

Unfortunately, activists use inaccurate characterizations to make it extremely time consuming and expensive to build new plants. Nuclear power is radioactive, they say—not mentioning that so is the sun and that taking a walk, let alone an airplane ride, exposes you to far more radioactivity than does living next to a nuclear power plant.[34] A nuclear plant could be bombed by terrorists and bring about some sort of Hiroshima 2, they say—not mentioning that the type of uranium used in a nuclear plant literally can't explode.

All of these fears are plausible because we have been taught to think of changing our environment in new ways as inherently dangerous. Nuclear power, in addition to requiring large industrial structures, deals in "unnatural" high-energy, radioactive materials and processes. Thus there is an *expectation* that it is uniquely dangerous, even though it is uniquely safe.

The opposition has led nuclear power to be considered far more dangerous than other sources, unjustifiably. And it means that the nuclear industry has become an essentially government-controlled industry—which, like many a government-controlled industry, has higher prices than others. Thus we don't really know what nuclear would cost without the pseudoscientific opposition. What we do know is that, besides fossil fuel energy, it is by far the most *scalable* form of energy in the world.

In the best-case scenario, though, nuclear is still decades away from scaling to becoming a leading global source of electricity, let alone somehow providing transportation solutions at the level oil can. Thus there is no prospect of nuclear "replacing" fossil fuels anytime soon.

3

THE GREATEST ENERGY TECHNOLOGY OF ALL TIME

FOSSIL FUEL POWER: CHEAP, PLENTIFUL, RELIABLE, SCALABLE—INDISPENSABLE

This is the challenge: finding a source of energy that is cheap, plentiful, reliable, and scalable. As we've seen, it's a challenge that is incredibly difficult to overcome. Power from sunlight has the problems of diluteness and intermittency and so requires too many resources to concentrate and store in order to create an independent, scalable power source. And plants are a form of storing solar energy, but they don't scale well because of the resources needed to grow them and the amount of land available to grow them on.

Well, there is good news. There is a form of solar energy, a biofuel that has none of these problems because nature has already concentrated and stored the sunlight of plants that lived hundreds of millions of years ago. Those dead plants are called fossil fuels.

Fossil fuels are so called because they are (in most theories) high-energy concentrations of ancient dead plants. Our entire civilization is based on burning these dead plants, which are made up of hydrogen and carbon atoms connected by powerful chemical bonds. When you burn gasoline in your car or coal in a power plant or gas to heat your home, those bonds break apart, releasing enormous amounts of energy. They exist in solid (coal), liquid (oil), and gas (natural gas) form.

If you've ever used charcoal instead of wood to grill food, you grasp the basic advantage of using ancient dead plant fuel. The charcoal can generate more heat in less space because it has been "cooked"—primarily, the water has been taken out of it, producing a higher concentration of energy ("burning" water doesn't release much energy).[1] Well, fossil fuels are naturally, thoroughly "cooked" plant energy. Over millions of years, as plants pile up and are covered by more and more layers of soil, the natural forces of the Earth heat them up and concentrate them into far purer forms of energy than wood or charcoal. Thus they have the advantage of being naturally concentrated and stored.

The other advantage they have is that they exist in astonishingly, astonishingly large quantities. For example, the world has an estimated 3,050 years (at current usage rates) of "total remaining recoverable reserves" of coal.[2]

But there is a big challenge to using fossil fuels for energy. These quantities of coal, oil, and gas aren't lying around to be plucked. They are *hidden and trapped* underground—sometimes thousands and thousands of feet underground, often in forms, such as being trapped in stone, that are difficult to get out even if you know where they are.

Fortunately for us, the fossil fuel industry is very, very good at using *technology* to extract these hidden, trapped, and otherwise useless materials, which no one knew about or cared about through most of human history, and turning them into the energy of life.

The technical term for fossil fuels is *hydrocarbons,* because they are primarily made of carbon and hydrogen atoms. Also, there is some debate over whether all of them come from plants (fossils); some say that many or most of them come from deep in the Earth, far below where any plants could end up. In either case, there are astonishing quantities of hydrocarbons. With ever-evolving technology, they give us an unparalleled source of concentrated, stored, and scalable energy.

Fossil fuels come in three major forms—coal, oil, and natural gas—with different strengths and weaknesses.

COAL

Coal is the world's leading fuel for electricity—producing 41 percent of the world's electricity in 2011—and is expected to become the leading source of energy overall.[3] In the developing world, it has been the overwhelming choice for every country that has industrialized recently.

Since the 1980s, the world has experienced record increases in coal consumption: in Brazil, by 144 percent; in India, by 425 percent; in China, by 514 percent.[4] It is no coincidence that countries with increased coal consumption also experience better lives overall—as electricity consumption increases, infant mortality rate decreases rapidly and access to improved drinking water sources increases.[5]

The reason coal is particularly well suited for cheap electricity around the world is that it is plentiful, widely distributed, and relatively easy to extract. Coal is also relatively easy to transport. It exists in a convenient form, and unlike most mine products, which require you to separate large amounts of material from the small amount of material you want, coal requires relatively little processing.

But because of its plant origins and underground locations, some of coal's carbon and hydrogen are bonded to potentially significant quantities of sulfur and nitrogen. When burned, these become sulfur dioxide and various nitrogen oxides, which above certain concentrations can be harmful, requiring various filtration and dilution technologies (more on this in chapter 6). Coal has the highest percentage of carbon atoms of all the fossil fuels, so when burned, it emits the most carbon dioxide, whose impact we will examine in chapter 4.

Coal has been used for transportation fuel and was the dominant form of energy for locomotives and steamships when the steam engine was still the main source of motive power.[6] Eventually the steam engine was supplanted by the much more versatile internal combustion engine, which has nearly eliminated coal's use as a transportation fuel in favor of oil. Because the fossil fuels' value comes from their being hydrocarbons—combinations of hydrogen and carbon—each of them can be made to have many of the properties of the others, but that transformation requires energy and resources, like any transformation, and is not often worth it. But in the future, it might be worth it—which means that claims that we'll "run out of oil" are misguided, as coal and gas can effectively produce oil if needed.

For example, coal can be transformed into liquid fuel; the South African energy company SASOL says it can be done for less than eighty dollars per barrel.[7] Coal can also be transformed into methanol—methyl alcohol, an alcohol that like ethanol can come from plants but can also come from coal and gas. Methanol, like any fuel, has its own risks and by-products, and it has only half the energy per gallon as gasoline, but it is still a potential substitute for oil fuel as markets evolve.

Coal use is growing quickly and could grow even more quickly. The United States could be a major contributor; we have been called the Saudi Arabia of coal and have the potential to become a huge coal exporter, feeding cheap energy to machines around the world.

The bottom line: If people are free to use it and the industry is free to produce it, coal energy will provide billions with cheap, plentiful, reliable energy for decades to come.

NATURAL GAS

Natural gas is the world leader at an essential type of electricity—called *peak load* electricity.

Just as your energy use varies during the course of a day, so the electric grid as a whole uses different amounts of electricity at various times during the day. There is a minimum amount of electricity use that will almost always be needed, called base-load power. Above that, we need a technology that can quickly adjust to changes in electricity needs—such as powering a lot of air conditioners on a hot summer day so that we can be comfortable and avoid heatstroke. This is called peak load electricity, and it is natural gas's specialty. (Coal, nuclear, and hydro specialize in base-load power.) Natural gas electricity, which uses the same basic technology as a jet engine, is very good at scaling up and down.

Natural gas is also an extremely clean-burning fuel, composed almost exclusively of pure carbon and hydrogen, which makes it ideal to burn for affordable home heating. In addition, it serves as an affordable, abundant raw material for thousands of "petroleum products"—which we will discuss in the next section.

The disadvantage of natural gas is in its name—it's naturally a gas. Gases are harder to move long distances than liquids or solids due to their large volume. So while oil and coal can be moved relatively easily around the world, gas has long been a local market. This causes supply security issues in which one country is dependent on an unreliable country for gas supplies—the case with many European countries that depend on gas from Russia.

However, new technological developments are overcoming these obstacles.

One is shale energy technology, often referred to as fracking in the media and fracing in the industry. Fracking is short for hydraulic fracturing, one of several technologies that can be used to get natural gas out of shale. This technology has attracted attention based on claims that it contaminates groundwater. As we'll discuss more in chapter 7, the controversy here, as with nuclear, is more ideological than technical.

The shale energy revolution has led to a rapid increase in natural gas and oil production in the last decade and has the potential to do much more.[8] The combination of horizontal drilling and fracking has turned previously known but economically unreachable reserves of natural gas into easily accessible and cheap natural gas. In the United States, proven reserves of natural gas have increased 46 percent since 2005.[9]

There are opportunities all around the world to produce shale energy, and the United States is a pioneer. There are estimated to be far more natural gas supplies in what are called methane hydrates, natural gas deposits in frozen form, which exist at the bottom of the ocean.[10] Thus the potential supply of natural gas could extend many centuries, at least.

At the same time, advances in compressing and liquefying natural gas are making it more prominent as a fuel and make it easier to transport around the world. This is the source of opportunities and controversies for LNG (liquefied natural gas) terminals.

Gas can also be turned into fuel oil and methanol, and it powers vehicles in compressed or liquefied form. Still, when it comes to transportation, nothing can yet compete with what is far and away the world's leading transportation fuel: oil.

OIL

Oil is the most coveted (and controversial) fuel in the world because it is almost eerily engineered by natural processes, not just for cheapness, not just for reliability, not just for scalability, but also for another characteristic crucial to a functional civilization: *portability*.

Oil is an ultraconcentrated form of energy—liquid energy—so it's ideal for any moving vehicle. Every portable power source needs to carry its fuel with it, which means that size and weight are of paramount consideration. Oil, in effect, has the ultimate strength to weight ratio. A gallon of gasoline has 31,000 calories—the amount of energy you use in *fifteen days*. Oil can be refined into stable, potent liquid fuels—gasoline, diesel, and jet fuel.

Oil's dominance as *the* transportation fuel has gone hand in hand with the development of mobile engines: the gasoline engines in most cars, the diesel engines powering semitrucks and global shipping, and the jet engines powering aircraft all eat oil fuel.

Oil is used for the vast, vast majority of transportation—93 percent in the United States.[11] Other technologies struggle to mimic it.

Oil's value leads to continuous large investments in exploration and extraction technology. Whereas oil deposits were once completely invisible to industry, today modern imaging, called 3D seismic imaging, can get us a far clearer idea of what's going on below the surface and how it changes over time. We can get oil out of hard rock (shale oil). In oil sands, we've created technology that acts as a ground decongestant—releasing oil from the sands that have held it in place for decades.

Portability is valuable for many reasons. Personally, oil is the fuel of freedom—the fuel that liberated Americans to go where they want, when they want. Economically, oil is the fuel of trade. Our entire standard of living depends on *specialization*—on people doing what they do best, wherever they are, and then being able to cheaply move their products to those who most need them. The higher the price of portable power, the slower the world economy moves.

In the future, it is quite possible that battery-powered vehicles will replace oil-powered vehicles for certain purposes. The limits are based on the energy concentration or energy density of the batteries; for example, the Tesla Roadster battery has an energy density that is 107 times *less* than gasoline—though the battery's electric motor can convert more than twice as much of that energy into usable energy as the engine in a gasoline-powered car, so in practice the Tesla battery might be 35 times less dense.[12]

For various technical reasons, progress in battery technology is extremely slow—electric cars have been around longer than gasoline-powered cars—and it may well be that another, nonbattery storage solution will win out. For now, though, oil is the greatest portable fuel the world has ever known, and we are willing to pay a premium for it; per unit of energy, we sometimes pay five times as much for oil as for natural gas.

Oil is also coveted as the world's most versatile raw material for making synthetic materials. You are probably sitting in a room with at least fifty things derived from oil, from the insulation in your wall to the carpet under your feet to the laminate on your table to the screen on your computer. Oil is everywhere—that is how the average American uses 2.5 gallons each day.[13]

Like coal and gas, there is enormous future potential for oil production—if the industry can keep developing better technologies. The shale energy revolution is bringing supplies some never expected, and the Earth still contains many times more oil than we have used in the entire history of civilization.

FUTURE ENERGY RESOURCES

Here's a trick question I like to ask when I do public speaking: "Is oil a valuable natural resource?" Almost everyone answers yes, even when I tell them it's a trick question.

My answer is no. Because oil—or coal, or natural gas, or uranium, or aluminum for that matter—is not naturally a resource.

If we understand this, we understand why we can be incredibly optimistic about the future potential of fossil fuels and future sources of energy.

A resource is something that's available and usable for human benefit. I'll focus on oil here because that is the resource that people most fear will disappear.

Before the 1850s, oil was not a resource—it was naturally useless. It was a distinct *raw material*, to be sure, with the potential to become valuable, just as sand has the potential to become a microchip. But oil had very little use; in fact, in many cases, it was a nuisance. Drillers seeking underground saltwater deposits to distill into salt were annoyed by the presence of this "rock oil."[14] Additionally, oil was not a resource because it was hidden and trapped, invisible and inaccessible.

What turned oil from a potential resource to an actual resource was human ingenuity—the ingenuity of the chemist Benjamin Silliman Jr., who refined petroleum into kerosene, the ingenuity of George Bissell, who targeted Titusville, Pennsylvania, as a location likely to have underground oil, and the ingenuity of Edwin Drake, who created the first successful oil well in 1859 at 69.5 feet underground.[15]

It was only thanks to their ingenuity that useless goo became a resource.

The history of oil is a history of *resource creation.* For example, crude oil, through a process of boiling (distilling), could be refined into 50 or 60 percent kerosene, used for lighting. But then the rest of the crude oil wasn't a resource—it was often pure waste, dumped in a lake—until human ingenuity made it so. In the nineteenth century, John D. Rockefeller's Standard Oil progressively figured out how to create value out of every "fraction" of a barrel—a barrel containing numerous types of hydrocarbons of different shapes, sizes, and masses. They created wax out of one part of the barrel, lubricants (over three hundred varieties) out of another, and asphalt out of another.

In the twentieth century, modern chemistry made oil not only the most important fuel, but also the most important raw material in civilization. Chemists can "crack"—break down—the molecules in a barrel of oil into small parts, and then reassemble them into an unbelievable variety of polymers, including modern plastics. While you think of oil in your car as in the gas tank, in fact there is more oil in the materials in the car than in the gas tank. The rubber tires are made of oil, the paint and waterproofing are made of oil, the plastic, dent-resistant bumper is made of oil, the stuffing inside the seats is made of oil, and in most cars, the entire interior is one form of oil fabric or synthetic material or another—because oil is such a cheap and effective way to make things.

When a policeman has his life saved by a bulletproof vest, when a firefighter has his life saved by a fireproof jacket, that is oil—that is something that was once a useless raw material, now made into a resource.

What is true of oil is true of essentially every other resource: They need to be *created* by transforming potential into actual. Coal was not an electricity resource or a source of motive power until the coal-fired steam engine. Natural gas was actually a deadly force, something that exploded when you drilled for oil, until safe drilling and storage technologies could harness it. Aluminum, one of the most abundant elements in the Earth's crust, was completely useless a few hundred years ago.

Ultimately, an "energy resource" is just matter and energy transformed to meet human needs. Well, the planet we live on is 100 percent matter and energy—100 percent potential resource. To say we've only scratched the surface is to significantly understate how little of this planet's potential we've unlocked. We already know that we have enough of a combination of fossil fuels and nuclear power to last thousands and thousands of years. For us today, that's morally enough—it's time to focus on the 7 billion of us, here and now, who will live better with more energy and live worse or not at all with less.

What energy resources should we use now and in the future? We have a brilliant system for deciding this: the price system of supply and demand. All things being equal, if it takes fewer resources, including human time, to produce something, the price goes down; if it takes more resources, the price goes up.

Thus, prices reflect how *efficient* a use of existing resources it is to create a new resource for a given purpose. When the cost of computers comes down, that means that all the components and their composition can be created more cheaply than before. Similarly, the form of energy we use will be the one that, based on the best technology available (which is always evolving), can do the best job for the lowest price.

Every day, we make a choice. Is coal or oil or gas the best way of accomplishing a given goal—or is something else? For the last several hundred years, the answer to "What do we replace yesterday's fossil fuels with?" has been "New fossil fuels." As soon as that doesn't make sense (typically, when it becomes prohibitively expensive or when a better alternative is available), it won't happen.

Part of the process of resource evolution is that we will find new ways to get what is considered to be the same resource by more technically complicated means. This is often characterized negatively, with such expressions as "We've gotten all the easy oil, and now we're going after the dirtiest oil" or "We're scraping the bottom of the barrel."[16] (The expression "scraping the bottom of the barrel" comes from the phenomenon of the oil in a barrel existing in different fractions, from heavy to light. The heavy fractions sit at the bottom of the barrel, and the heaviest, like asphaltum, which goes into asphalt, can be hard to scrape out and impossible to use.)

The view is that when we use a finite, nonrenewable resource like fossil fuels, we will have to go to progressively more difficult places to get it—which is assumed to be a bad thing. But why? Every resource technology involves starting with easier problems and moving toward harder problems.

When I read "We're using dirtier and dirtier oil" or "We're having to scour further depths to get oil," I think, *What is the "appropriate'" length to go to get oil? Should we have stopped at 69.5 feet?* At every stage, one could be accused of "scraping the bottom of the barrel." But think forward two hundred years. The Earth is full of elements at the crust and below. Ditto for the bottom of the ocean. Someday we will likely have technology to mine the ocean floor more efficiently than we can mine at the surface today. What is wrong with going to that frontier? How is that any different than going to space?

Now, it makes no sense to go to such great lengths to get oil if it's not efficient, i.e., if there are better alternatives. But there will likely be some element that is most efficient to get from very far below the ground.

The most forward-looking policy toward energy use is to always use the most competitive form of energy. I like to call the most competitive ones *progressive energy,* because they are part of a process of continual improvement, of finding the best way to get energy from the Earth's effectively unlimited stockpile of potential energy resources.

Our concern for the future should not be running out of energy resources; it should be running out of the *freedom* to create energy resources, including our number-one energy resource today, fossil fuels.

EVERY CALORIE MATTERS

Because we have never lived without fossil fuel energy, it's hard to imagine life without its benefits. But given that thought leaders are proposing exactly that, it's important to grasp just how big a difference fossil fuels have made in our lives.

To get a big-picture view of the difference energy and machines make in our lives, look at this graph of human progress from A.D. 1 to the present, featuring data from the Angus Maddison survey, the most comprehensive survey of quality of life over the last two millennia.

Figure 3.1: Fossil Fuel Use and Human Progress—the Big Picture

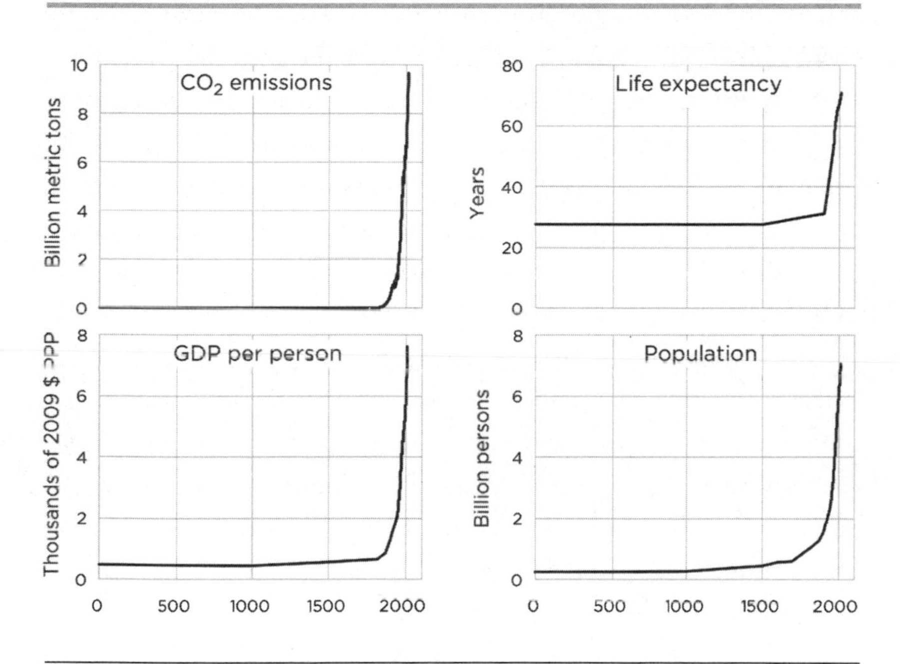

Sources: Boden, Marland, Andres (2010); Bolt and van Zanden (2013); World Bank, World Development Indicators (WDI) Online Data, April 2014

Notice that starting in the year 1800, the metrics of life expectancy and GDP rocket up. In school, we learn that this was the time of the Industrial Revolution—although at least in my case, the problems of it were emphasized much more than the doubling of human life expectancy and the far more than doubling of individual income. What exactly does the term *industrial revolution* mean? Well, it was a revolution in industry, which means in our ability to do physical work to be more productive, which in practice meant an *energy* revolution. Thanks to the world's first source of cheap, plentiful, reliable energy, coal-powered steam engine technology, every industry became more productive—agriculture, manufacturing, transportation.

The people who went through the industrial revolution had a perspective that is hard for us to recapture but is essential for us to get: an understanding of just how vital it is for us to have access to cheap, plentiful, reliable energy, because the more we have, the more ability we have, and the less we have, the more we see just how weak we are without high-energy machines. I stress "cheap, plentiful, and reliable" because anything less than that isn't useful, just like the expensive, scarce, unreliable electricity at the Gambian hospital. Before the industrial revolution, there were machines and there were sources of energy—there just wasn't cheap, plentiful, reliable energy for the vast majority of people.

Take this 1865 comment by William Stanley Jevons, a legendary theoretical and applied economist, in his book *The Coal Question*. Coal was then the cheapest and most reliable source of energy, not only for illumination, which is only one use of energy, but also for powering machines to do far more mechanical work than human beings can with our muscles.

> Coal in truth stands not beside but entirely above all other commodities. It is the material energy of the country—the universal aid—the factor in everything we do . . . new applications of coal are of an unlimited character. In the command of force, mo-

> lecular and mechanical, we have the key to all the infinite varieties of change in place or kind of which nature is capable. No chemical or mechanical operation, perhaps, is quite impossible to us, and invention consists in discovering those which are useful and commercially practicable. . . .[17]

Jevons was worried that we were running out of coal (a concern we'll discuss later in this chapter). Notice how emotional he is about it:

> With coal almost any feat is possible or easy; without it we are thrown back into the laborious poverty of early times.[18]

A letter in response to Jevons is even more vivid about what the loss of coal energy would have meant:

> Coal is everything to us. Without coal, our factories will become idle, our foundries and workshops be still as the grave; the locomotive will rust in the shed, and the rail be buried in the weeds. Our streets will be dark, our houses uninhabitable. Our rivers will forget the paddlewheel, and we shall again be separated by days from France, by months from the United States. The post will lengthen its periods and protract its dates. A thousand special arts and manufacturers, one by one, then in a crowd, will fly the empty soil, as boon companions are said to disappear when the cask is dry.[19]

And here's how it ends: "We shall miss our grand dependence, as a man misses his companion, his fortune, or a limb, every hour and at every turn reminded of the irreparable loss."[20]

One thing to note: This was at a time in history when, because of early-stage technology, coal pollution in England was far, far worse than, say, even China experiences today—and yet these commentators don't even mention it; that's how valuable they saw energy as being to their very ability to survive. Nothing was more important. As we'll see looking at modern fossil fuel technology, we have progressed incredibly in pollution-reduction technology, but it's worth remembering that to the people who experience the need for energy most directly, it's worth pretty much any price, in the same way that you'll put up with a lot of side effects to take a lifesaving drug. And lifesaving drugs, like everything else we value, depend on access to cheap, plentiful, reliable energy—to produce, to transport, to package, to refrigerate.

When we talk about different sources of energy, we are talking about different technologies that are better or worse at producing energy with the resources we have. If we choose the most capable technologies, we get more energy. If we choose less capable ones, we get less. It's that simple. When someone says, "Let's use solar," he is, usually unwittingly, saying, "Let's have less energy with which to improve our lives." There is no limit to the amount of energy we can use to improve our lives. And in a world where we produce only one fourth as much energy as would be necessary for everyone to live like Americans, every machine calorie counts.

One realm in which energy is particularly life or death is in agriculture. The fossil fuel industry has revolutionized acriculture to the benefit of billions—and gotten no credit.

MORE FOSSIL FUELS, MORE FOOD: HOW THE OIL INDUSTRY SOLVED WORLD HUNGER

Paul Ehrlich declared in the 1968 sensation *The Population Bomb* that "the battle to feed humanity is over"—and he was in good company.[21] In 1969, the *New York Times* reported: "While there have always been famines and warnings of famine, food experts generally agree that the situation now is substantially different. The problem is becoming so acute that every nation, institution, and every human being will ultimately be affected."[22] A group of leading American intellectuals wrote an open letter declaring: "The world as we know it will likely be ruined before the year 2000. . . . World food production cannot keep pace with the galloping growth of population."[23]

In 1968, the world's population was 3.6 billion people.[24] Since then, it has doubled, yet the average person is *better fed* than he was in 1968.[25] This seeming miracle was due to a combination of the fossil fuel industry and genetic science—such as the achievements of the great Norman Borlaug, who bred new revolutionary wheat varieties and introduced new farming techniques to Mexico, India, Pakistan, China, and parts of South America.

Modern agriculture, like every modern industry, runs on machines, and fossil fuel energy is our leading source of machine food. Therefore, fossil fuel energy is the food of food.

For example, oil-powered *mechanization* causes a dramatic increase in the amount of farmland that can be cultivated per worker.

For most of human history, agricultural work was done by the muscle power of humans or draft animals, placing a low ceiling on the amount of farmland that could be harvested—and requiring often 90 percent of populations to be devoted to farm labor. The oil industry changed that by making available cheap, concentrated energy that could power tractors, combines, and other forms of high-powered farm equipment. Matt Ridley, author of the valuable survey of human progress, *The Rational Optimist,* describes the value of mechanization on his own farm: "A modern combine harvester, driven by a single man, can reap enough wheat in a single day to make half a million loaves."[26] A single man, made into an agricultural Superman by the power of oil.

Another example: Oil-based *transportation* causes a dramatic increase in the amount of farm products that can be brought to market.

For the vast majority of human history, the world was full of patches of useless potential farmland—useless because the land was too far to ship from. When men and goods travel by horse or mule, let alone on foot, the shipping costs quickly exceed the value of the cargo. But the twentieth century's gradual increase in oil-powered transportation—railroads (modern railroads are powered by diesel engines), freighters, and trucks, especially—brought an enormous amount of remote farmland, once too expensive to ship from, within the reach of anyone in the city, state, country, and eventually the world. The cheaper transportation became, the more farmland came into the global agricultural economy, and the more plentiful and affordable food became.

By the same token, the cheaper transportation became, the more new seeds and other supplies could be brought to new locations to make previously low-performing land yield a giant amount of crops. Much of the green revolution led by Norman Borlaug involved bringing in new, more resilient forms of wheat and rice to places like India; this was expedited and amplified by cheap, global, oil-powered transportation.

Another example: Gas-based *fertilization* increases the amount of crops that can be grown per unit of farmland.

The amount of crops we can grow today is an utterly "unnatural" phenomenon—that is, it is way beyond the natural capacity of the nutrients in land to nourish crops in one season, let alone season after season. One solution to the problem of fertilizing was manure or some other organic fertilizer, which increased the amount of nitrogen plants could absorb and thus the amount of them that could grow. The use of such fertilizer allowed population growth and living standards to rise throughout the nineteenth century. But there was a problem; as population grew, it was harder to find enough manure to collect. The supplies of guano off the coasts of South America and South Africa were being exhausted, which caused eminent chemist William Crookes to declare in 1898 that "all civilisations stand in deadly peril of not having enough to eat."[27]

The solution was Fritz Haber and Carl Bosch's process of making large quantities of synthetic nitrogen fertilizer using enormous amounts of methane—the predominant component of natural gas.[28]

Another example: Electricity-based (usually coal-based) or diesel-based *irrigation* increases the amount and reliability of water going to crops. Irrigated lands average more than *three times* the crop yields of rain-fed areas. Sometimes irrigation occurs via gravity, but when it doesn't, it takes a lot of energy—usually fossil fuel energy—to move the water.[29]

Finally, the achievements of Norman Borlaug and other great food scientists, often called the green revolution (not related to the modern Green movement), were possible only because of the *time* created by fossil-fueled civilization to engage in intensive research, because high-powered machines have made it unnecessary for all of us to do physical labor.

Fossil fuel energy is the food of food.

It is an undeniable truth that, in providing the fuel that makes modern, industrialized, globalized, fertilized agriculture possible, the oil industry has sustained and improved billions and billions of lives. If we rate achievements by their contribution to human well-being, surely this must rank as one of the great achievements of our time, and when we consider the problems with that industry, shouldn't we take into account that it fed and feeds the world? And yet have you ever—and I mean ever—heard any major public or private figure give the oil industry credit for it? I see Bono and other celebrity activists get credit for caring but not the oil and energy industries for *doing*.

MORE FOSSIL FUELS, MORE ABILITY

Without the energy industry, the agricultural industry would not exist; the world could not support a population of 7 billion or 3.6 billion and perhaps not even 1 billion. To starve our machines of energy would be to starve ourselves.

What is true of agriculture is true of every industry. The energy industry has a special place in human productivity, prosperity, and progress. As the industry that powers every other industry, it can be considered *the master industry*. Whether we are talking about the computer industry, the electronics industry, the health-care industry, or the pharmaceutical industry, every industry uses machines, uses resources that are manufactured using energy, and uses *time* that is available because of our high-energy society's productivity. The less energy we have, the fewer machines an industry can use, the fewer resources an industry has, and the less time it has. And what happens to industry happens to the rest of life. The less productive industry is, the less time, resources, and machinery we have to enjoy our lives.

And I want to stress *enjoy our lives,* because this is not something we typically think about when we think of energy—but we should, because more energy means more ability to enjoy our lives.

If our standard of value is human life, the ultimate benefit that a commodity like fossil fuel energy can deliver is to contribute to the pursuit of happiness. If we can only survive in a way that is miserable, why survive? Happiness is the reward of life. And energy is a great enabler of happiness—including forms of happiness that we are taught to associate with people who decry large amounts of energy use.

Take traveling to places that excite us. In my life, I have been fortunate enough to travel to many such places. I've spent fifteen days river-rafting in the Grand Canyon, several hundred days snowboarding or skiing on faraway mountains, and gone to Italy, France,

Israel, Turkey, and other faraway lands. I'm not setting any travel records here, but no doubt I've used a lot of cheap energy to enjoy what the world has to offer. On a more local level, I love living in Southern California and being able to get to a lot of places easily by car (assuming I time the traffic properly). I enjoy martial arts, specifically Brazilian Jiu-Jitsu, as a hobby, and for years I would happily drive an hour after work almost every day to get to my favorite Jiu-Jitsu school. I was using a lot of cheap energy, and if it hadn't been cheap, I wouldn't have been able to afford it.

More fossil fuel energy, more ability to pursue happiness.

I keep stressing that more energy means more ability to take the actions necessary to flourish, because I want it to be in our minds at all times that when we talk about more or less energy, we are talking about more or less ability, and everything we want in life depends on ability. Thus in every realm that affects our lives, we should expect to discover that more energy can play an amazingly positive role.

That includes our *environment*, including our climate.

FOSSIL FUEL ENERGY AND OUR ENVIRONMENT

The relationship between energy and environment is usually considered in a negative way; how can we use the energy that will least "impact the environment"? But we have to be careful; if we're on a human standard of value, we *need* to have an impact on our environment. Transforming our environment is how we survive. Every animal survives in a way that affects its environment; we just do it on a greater scale with far greater ability. We have to be clear: Is human life our standard of value or is "lack of impact" our standard of value?

If we're on a human standard, we should be concerned in a negative way only about impacts of energy use that harm our environment *from a human perspective*—such as dumping toxic waste in a nearby river or filling a city with smog.

But we should also assume that energy gives us more ability to *improve* our environment, to make it healthier and safer for human beings. I'll explore this in detail in chapters 4–7, but for now I'll just observe that the natural environment is not naturally a healthy, safe place; that's why human beings historically had a life expectancy of thirty. Absent human action, our natural environment threatens us with organisms eager to kill us and natural forces, including natural climate dangers, that can easily overwhelm us.

It is only thanks to cheap, plentiful, reliable energy that we live in an environment where the water we drink and the food we eat will not make us sick and where we can cope with the often hostile climate of Mother Nature. Energy is what we need to build sturdy homes, to purify water, to produce huge amounts of fresh food, to generate heat and air-conditioning, to irrigate deserts, to dry malaria-infested swamps, to build hospitals, and to manufacture pharmaceuticals, among many other things. And those of us who enjoy exploring the rest of nature should never forget that energy is what enables us to explore to our heart's content, which preindustrial people didn't have the time, wealth, energy, or technology to do.

We'll revisit this topic later; for now, I just want to stress that whenever we have more energy, we have more ability *everywhere*—including the places we can do damage. So when we look at the damage or the risks of damage, we have to take into account the positive as well. Once again, we're always looking for the big picture about what benefits human life.

THE BIG PICTURE

We have seen that the non–fossil fuel attempts at cheap, plentiful, reliable energy for billions of people fall short—because none of them involve a process wherein *every element* can be scaled cheaply and reliably.

But fossil fuel technology puts everything together: It can get a plentiful fuel source cheaply and convert it to energy cheaply—on a scale that can power life for billions of people. This is why when people choose to use energy to improve their lives, 87 percent of the time they choose fossil fuel energy.[30] The technology is that far ahead of the competition. If we want cheap, plentiful, reliable energy around the globe, we absolutely need to use fossil fuel technology. If we want to flourish, we need fossil fuel technology.

And yet opponents of fossil fuel energy claim there are catastrophic consequences to using fossil fuels that will *prevent* us from flourishing. That will be our subject for the next several chapters.

But before we get there, let's be clear: *If* fossil fuels have catastrophic consequences and it makes sense to use a lot less of them, that would be an epic tragedy, given the state of the alternatives right now. Being forced to rely on solar, wind, and biofuels would be a horror beyond anything we can imagine, as a civilization that runs on cheap, plentiful, reliable energy would see its machines dead, its productivity destroyed, its resources disappearing.

Thus it is disturbing to hear politicians talk about restricting fossil fuels as an "exciting opportunity." John Kerry, our secretary of state, whose job is to represent the mainstream views of America to the rest of the world, described the prospect of outlawing the vast majority of fossil fuels, even if there were no catastrophic climate change, this way:

> If the worst-case scenario about climate change, all the worst predictions, if they never materialize, what will be the harm that is done from having made the decision to respond to it? We would actually leave our air cleaner. We would leave our water cleaner. We would actually make our food supply more secure. Our populations would be healthier because of fewer particulates of pollution in the air—less cost to health care. Those are the things that would happen if we happen to be wrong and we responded.[31]

Actually, the type of "response" governments around the world have embraced—an 80 percent reduction in CO_2 emissions over several decades—would, by all the evidence we have, lead to billions of premature deaths.

Fossil fuel energy is, for the foreseeable future, necessary to life. The more of it we produce, the more people will have the ability to improve their lives. The less of it we produce, the more preventable suffering and death will exist. To not use fossil fuels, therefore, is beyond a risk—it is certain mortal peril for mankind.

That brings us to the issue of the major risks cited with fossil-fuel use: climate change and environmental degradation. As we begin to think about risks, we need to keep this in mind: The reason we care about risk is because it is a danger to human life. Thus if something is essential to human life, like fossil fuels, we need to assess all risks in that context.

We need a rigorous, big-picture examination of fossil fuels' impact on climate and other environmental issues. We must clearly hold human life as our standard of value, or if we don't, we must make clear that we are willing to sacrifice human life for something we think is more important. With that standard, we must look at the big picture, the full context. And we must use experts as advisers, not authorities, getting precise explanations from them about what is known and what is not known, so that we as individuals can make the most informed decision.

4

THE GREENHOUSE EFFECT AND THE FERTILIZER EFFECT

CLIMATE CONFUSION

Growing up in Chevy Chase, Maryland, a suburb inside the Beltway of the D.C. metro area, I learned only one thing about fossil fuels in school for the first eighteen years of my life: They were bad because they were causing global warming. It wasn't very clear in my mind what warming was or how it worked, but the gist was this: The CO_2 my parents' SUV was spewing in the air was making the Earth a lot hotter, and that would make a lot of things worse. Oh, and there was one more thing I learned: that everyone who knew the relevant science agreed with this.

Perhaps this would make a better story if I told you that I promptly joined Greenpeace and fought fossil fuels until discovering a massive hoax that I will reveal later in this chapter.

But that's not quite how it went. As a young free-marketer, my sixteen-year-old self did not like all the talk of political restrictions that went along with global warming. So I wasn't going anywhere near Greenpeace. But at the same time, the idea that this was a matter of established science was extremely significant to me. I come from a family of scientists (two of my grandparents were physicists, two were chemists) and I was being told about global warming not by scientifically illiterate teachers who repeated what they read in the paper (well, not only by those), but by my math and science teachers at the internationally renowned Math, Science, and Computer Science Magnet Program at Montgomery Blair High School.

My strongest memory from my senior year statistics class is of the time when my teacher, a very bright woman, stopped talking about statistics one day and started talking about the perils of global warming. That she brought it up in statistics class and that she was so adamant about it gave all of us the impression that this was an issue the scientifically minded should get involved with.

It was the same story at Duke. In freshman chemistry, local legend teacher James Bonk explained that the greenhouse effect was simple physics and chemistry and denounced the Republicans who denied it.

At that time, as I went searching for alternative views, I became familiar with the existence of professionals in climate science, such as Richard Lindzen of MIT and Patrick Michaels of the University of Virginia, who argued that global warming wasn't the big deal it was made out to be.[1]

What was I supposed to make of all this? Should I go by the more popular position? My science teachers had taught me that this, historically, was a recipe for failure, and that we should believe things only if someone can give compelling evidence for them.

But there was so much going on in discussions of global warming, I didn't know how to decide where the evidence lay. I would hear different sides say different things about sea levels, polar bears, wildfires, droughts, hurricanes, temperature increases, what was and wasn't caused by global warming, and on and on.

With such a mess to work with, I—like most, I think—tended to side with the scientists or commentators whose conclusions were more congenial to me. I will admit to reiterating the arguments of skeptics of catastrophic global warming with the lack of rigor I think is extremely common among believers. But I didn't do this for long. I acknowledged that I didn't really know what to think, and the idea that we might be making the Earth fundamentally uninhabitable scared me.

CLIMATE CLARITY

My greatest moments of clarity came whenever I discovered an author or speaker who, instead of giving his particular answer to the question of global warming, would *try to clarify the questions*. For example: "What exactly does it mean to believe in 'global warming'?" Some warming or a lot? Little deal or big deal? A little man-made or a lot man-made? Accelerating or decelerating?

Having a background in philosophy, I recognized that most discussion of global warming would not stand up to fifteen seconds of scrutiny by Socrates, who alienated fellow Athenians by asking them to define what they meant when they used terms vaguely. I think Socrates would have been all over anyone who spoke vaguely of global warming or climate change without making clear which *version* of that theory they meant: mild warming or catastrophic warming.

A huge source of confusion in our public discussion is the separation of people (including scientists) into "climate change believers" and "climate change deniers"—the latter a not-so-subtle comparison to Holocaust deniers. "Deniers" are ridiculed for denying the existence of the greenhouse effect, an effect by which certain molecules, including CO_2, take infrared light waves that the Earth reflects back toward space and then reflect them back toward the Earth, creating a warming effect. But this is a straw man. Every "climate change denier" I know of recognizes the existence of the greenhouse effect, and many if not most think man has had some noticeable impact on climate. What they deny is that there is evidence of a *catastrophic* impact from CO_2's warming effect. That is, they are expressing a different opinion about how fossil fuels affect climate—particularly about the nature and magnitude of their impact.

Once I was clear on how unclear the questions we were asking were, I could ask better questions and get better answers. And once I got clearer on how to use experts as advisers, not authorities, and how to always keep in mind the big picture, I had a much better chance of getting the right answers to the right questions.

Here's how I put the right questions now, from a human standard of value.

The first is: How does fossil fuel use affect *climate livability*? When we burn fossil fuels, what are all the climate-related risks *and all the benefits* that result?

Given that our standard is human life—we want the climate we live in to be as livable as possible—there are two types of impacts we need to study and weigh. The first is the impact of CO_2 on climate itself. CO_2 affects climate in at least two ways: as a greenhouse gas with a warming impact, but also as plant food with a fertilizing impact (plants are a major part of the climate system as well as a benefit of a livable climate). I'll refer to these as the greenhouse effect and the fertilizer effect. The second impact of CO_2, which is rarely mentioned, is the tendency of cheap, plentiful, reliable energy from fossil fuels to amplify our *ability to adapt to climate*—to maximize the benefits we get from good weather and ample rainfall and minimize the risks from heat waves, cold snaps, and droughts. I'll refer to this as the energy effect.

Discussion of climate change often assumes that any man-made climate change is large if not catastrophic and that our ability to adapt is not all that important. This is unacceptable. It is prejudicial to assume that anything is big or small, positive or negative, before we see the evidence. We have to actually investigate the facts. It might be that the greenhouse effect leads to a tiny, beneficial amount of warming or that having or not having fossil fuels to build sturdy infrastructure is the difference between two hundred and two hundred thousand people dying in a hurricane.

Granted, acquiring evidence is often hard because of so many conflicting reports, which is why it's so important to get experts to explain what they know *and what they don't know* clearly and precisely.

The bottom line: For the three major climate impacts of fossil fuels—the greenhouse, fertilizer, and energy effects—we want to know how they work and how they affect us, all the while asking, "How do we know?"

CLIMATE LIVABILITY 101

To understand how each climate-related effect of fossil fuels works, we need to be clear on what exactly we're talking about when we talk about climate and climate livability. And a good place to begin is with the atmosphere.

The atmosphere is the mixture of gases around the Earth (held by its gravitational field) that makes life possible with oxygen (that humans breathe), carbon dioxide (that plants breathe), nitrogen (that plants eat), et cetera. It is a fascinating, fluid system that causes the heat of the sun and the water of the oceans and the plant life on the Earth's surface to lead to all kinds of local weather conditions around the globe.

Weather refers to present, near-term atmospheric conditions, especially temperatures and precipitation. At any given time on Earth, there exists a huge range of colder and warmer climates with different weather patterns that have different benefits and risks for human life. *Climate* is the longer-term (usually measured in thirty-year increments) weather trends in a given region: how hot and cold it gets, how much precipitation there is, what kind of storms pop up, et cetera. The global climate system is the sum of atmospheric conditions around the globe over time.

Talk about "the climate" tends to misrepresent how climate works. It makes climate seem like something uniform and unchanging rather than one part of a diverse, ever-changing system.

Climate change is a change in the general weather patterns on a local level. Global climate change, often equated with climate change per se or man-made climate change, is change in the overall climate system and its diverse subclimates. There are many factors that affect local and global climate, including changes in the sun's intensity, and changes in plant life that alter the concentration of different elements of the atmosphere and thereby change, for example, the amount of water vapor in the air. Locally, human activity can have major impacts. In Phoenix, for example, temperatures in the center of the city are up to 10 degrees Fahrenheit higher than in the rural areas.[2]

How can climate and climate change affect us? One crucial truth is that climate is naturally volatile and dangerous. Absent a modern, developed civilization, any climate will frequently overwhelm human beings with climate-related risks—extreme heat, extreme cold, storms, floods—or underwhelm human beings with climate-related benefits (insufficient rainfall, insufficient warmth). Primitive peoples prayed so fervently to climate gods because they were almost totally at the mercy of the naturally volatile, dangerous climate system.

In any era, it's easy to think that volatile, dangerous weather is unique to our era and must prove some dramatic climate change, whether natural or man-made. Every year, the news is full of headlines about dramatic, often tragic climate-related events—headlines like these:

- "20,000 Killed by Earthquake: Toll Is Growing, Bodies Float Down Ganges to the Sea"[3]

- "100 Are Injured, Property Damage Exceeds $1,000,000: Tornado Strikes Three States, Bitter Cold in North Area"[4]
- "Death's Toll Mounts to 60 in U.S. Storms"[5]
- "1,500 Japanese Die in Hakodate Fire; 200,000 Homeless: Largest City North of Tokyo Is in Ruins and Mayor Says It Is 'a Living Hell'"[6]
- "Where Tidal Wave Ruined Norway Fishing Towns"[7]
- "Antarctic Heat Wave: Explorers Puzzled but Pleased"[8]
- "7 Lives Lost as Tropical Storm Whips Louisiana: Hurricane Moves Far Inland Before Blowing Out Its Wrath in Squalls"[9]
- "Widely Separated Regions of the Globe Feel Heavy Quake"[10]
- "Earth Growing Warmer: What Swiss Glaciers Reveal"[11]
- "Death, Suffering over Wide Area in China Drouth [Drought]"[12]
- "Toll of Flood at High Figure: Over 100 Bodies Recovered and 500 Persons Missing in Southern Poland"[13]
- "Cuban Malaria Increases: Thousands Become Ill in Usual Seasonal Spread of Disease"[14]
- "Mid-West Hopes for Relief from Heat; 602 Killed"[15]
- "Famine Faces 5,000,000 in Drouth [Drought] Area"[16]
- "Rumanians Are Alarmed by Epidemic of Cholera"[17]

While these headlines read like they're straight out of today's news, they are actually from 1934—before significant CO_2 emissions began. Climate is always volatile, climate is always dangerous.

Or take the issue of sea levels. We are taught to think of sea level rises as an evil inflicted on nature by human CO_2 emissions. We will explore today's sea level trends, and the role of fossil fuels in them, later in this chapter, but it is almost universally conceded that any sea level rise today is tiny compared with the enormous, rapid sea level rises that have occurred over the last ten thousand years.

Thus, climate change, extreme weather, volatility, and danger are all *inherent in climate whether or not we affect it with CO_2 emissions.*

Thus, when we think about how fossil fuel use impacts climate livability, we are not asking: Are we taking a stable, safe climate and making it dangerous? But: Are we making our volatile, dangerous climate safer or more dangerous?

We'll start with the potential source of risk: the greenhouse effect.

THE GREENHOUSE FEAR

The greenhouse effect is the centerpiece of the prediction of catastrophic climate change. There are basically three parts to the prediction. (1) Man-made greenhouse gases emitted by fossil fuel combustion will cause dramatic warming of the global climate system. (2) Dramatic global warming will cause a dramatic, harmful change in the global climate system. (3) Those changes will overwhelm human beings' capacity for adaptation, rendering the planet far less livable.

Those are the steps that lead numerous scientists, environmental leaders, and political leaders to make statements like that of James Hansen, probably the world's most politically prominent climate scientist: "CEOs of fossil energy companies know what they are doing [by emitting CO_2] and are aware of long-term consequences of continued business as usual. In my opinion, these CEOs should be tried for high crimes against humanity and nature."[18]

If any element of the greenhouse fear turns out to be false—if CO_2 emissions don't cause dramatic warming, if dramatic warming doesn't cause harmful climate change, or if human beings can adapt well, then CO_2 emissions are not catastrophic.

In investigating whether they are or not, we'll start with the foundation: the amount of warming caused by the greenhouse effect from adding more CO_2 to the atmosphere.

WHAT EXACTLY IS THE GREENHOUSE EFFECT?

The greenhouse effect is a warming effect that certain molecules, including water and carbon dioxide, have when they are in the atmosphere. When infrared radiation from the sun reflects off the planet and heads toward space, these molecules, called infrared absorbers, reflect some of it back, causing heat.[19] The impact of these gases in the atmosphere is analogized, in its warming impact, to the glass in a greenhouse that helps keep plants warm.

Thanks to the greenhouse effect, the surface of Earth is many degrees warmer than it would otherwise be. Many scientists say that without it, the planet would be some 33 degrees Celsius (59 degrees Fahrenheit) colder—an ultra–Ice Age.[20]

When fossil fuels—hydrocarbons—are burned, or *oxidized,* the hydrogen becomes H_2O and the carbon becomes CO_2.

It's worth noting that every part of this process has climate impacts. The H_2O introduces new water vapor into the climate system and the burning of fossil fuels adds heat to the system—but both of these impacts are too small to make a noticeable difference. Much more significant, the human activities powered by fossil fuels are perfectly capable of affecting local climates. In cities, the bricks, pavement, and buildings impede the flow of ventilating winds, raising temperatures, especially nighttime lows, making heat waves more frequent. This man-made local warming is often far greater than the global warming trend over the last 150 years, which is .8 degree Celsius (1.44 degrees Fahrenheit), a quantity that cannot be perceived without instruments).[21]

Now let's look at CO_2. It's a greenhouse gas that exists in trace quantities in the atmosphere—just under .03 percent (270 parts per million, or ppm) before the industrial revolution, a level that we have increased to .04 percent (396 ppm).[22]

How do we know about the greenhouse effect of CO_2? The best way: it can be studied in a laboratory. The temperature difference between a box with a glass ceiling and normal atmospheric gas concentrations and one with additional CO_2 is measured when sunlight shines into it.

As with any effect, a crucial question is: What is its magnitude—including, at what rate does additional CO_2 change the effect? Some phenomena are linear, which would mean that every molecule of CO_2 you add to the system will add a unit of heat the same size as the last one. In some phenomena, the effect is constantly increasing or accelerating; in this case, every molecule of CO_2 you add to the system would be more potent than the last (this is the sense that we get from most popular treatments of the greenhouse effect). Then there are diminishing or decelerating phenomena—every molecule of CO_2 you add to the system would be less potent than the last.

Anyone discussing this issue should know what kind of function the greenhouse effect follows. While I've met thousands of students who think the greenhouse effect of CO_2 is a mortal threat, I can't think of ten who could tell me what kind of effect it is. Even "experts" often don't know, particularly those of us who focus on the human-impact side of things. One internationally renowned scholar I spoke to recently was telling me about how disastrous the greenhouse effect was, and I asked her what kind of function it was. She didn't know. What I told her didn't give her pause, but I think it should have.

As the following illustration shows, the greenhouse effect of CO_2 is an *extreme diminishing effect*—a *logarithmically* decreasing effect.[23] This is how the function looks when measured in a laboratory.

Notice that we are just before 400 ppm (which means CO_2 is .04 percent of the atmosphere), where the effect really starts tapering off; the warming effect of each new molecule is not much.

This means that the initial parts per million of CO_2 do the vast majority of the warming of our atmosphere. The image below shows how, all things being equal, the heating effect of each additional increment of CO_2 is smaller and smaller.

Figure 4.1: The Decelerating, Logarithmic Greenhouse Effect

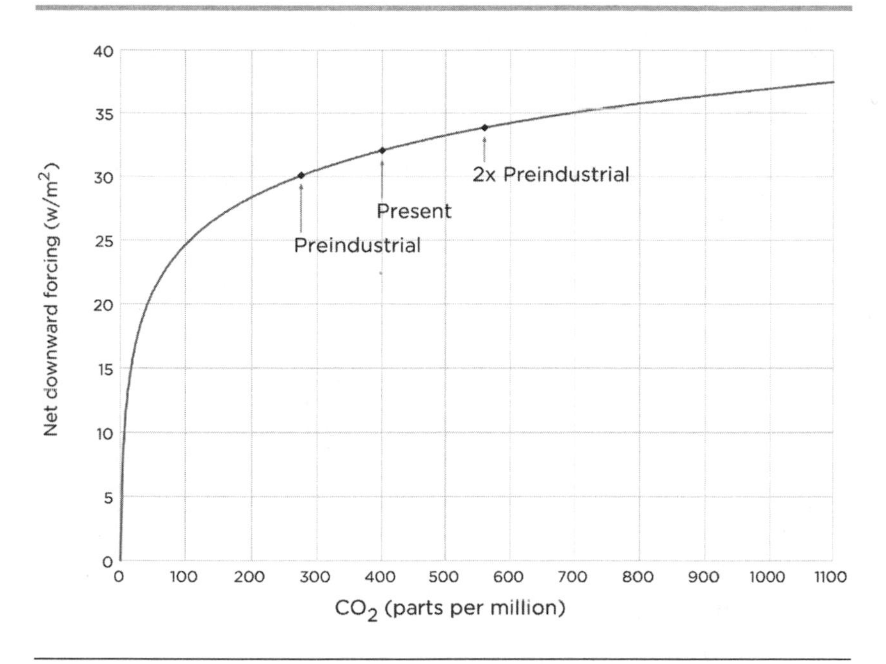

Source: Myhre et al. (1998)

So why do we have the idea that the greenhouse effect means rapid global warming? Because the proven greenhouse effect is *falsely equated* with the *related but speculative* theory that the greenhouse effect of CO_2 is dramatically *amplified* by other effects in the atmosphere, leading to rapid warming instead of the otherwise expected decelerating warming.

Some predictions of dramatic global warming (and ultimately catastrophic climate change) posit that the greenhouse effect of CO_2 in the atmosphere will greatly amplify water vapor creation in the atmosphere, which could cause much more warming than CO_2 acting alone would. This kind of reinforcing interaction is called a *positive feedback loop.*

What is the evidence for these predictions compared to the greenhouse effect?

To listen to most cultural discussion, predictions of dramatic global warming and associated catastrophic climate change are absolutely certain.

Secretary of State John Kerry said "absolutely certain" in a landmark speech discouraging the people of Indonesia from using fossil fuels, after they have experienced a major increase in prosperity due to increasing use of fossil fuels:

> The science of climate change is leaping out at us like a scene from a 3D movie. It's warning us; it's compelling us to act. And let there be no doubt in anybody's mind that the science is absolutely certain. It's something that we understand with absolute assurance of the veracity of that science. . . . Well, 97 percent of climate scientists have confirmed that climate change is happening and that human activity is responsible. These scientists agree on the causes of these changes and they agree on the potential effects . . . they agree that, if we continue to go down the same path that we are going down today, the world as we know it will change—and it will change dramatically for the worse.[24]

We'll get back to the 97 percent number in a minute, but if you press any climate scientist for an *explanation,* he will explain (or admit) to you that there is nothing resembling absolute certainty about these large positive feedback loops and the predictions associated with them. This is called the problem of determining *climate sensitivity;* how much warming, in practice, in the full complexity of the atmosphere, does *x* amount of CO_2 cause? How strong a *driver* of climate is CO_2?

Those who speculate that CO_2 is a major driver of climate have, to their credit, made predictions based on computer models that reflect their view of how the climate works. But fatally, those models have failed to make accurate predictions—not just a little, but completely.

While everyone acknowledges that the climate is too complex to predict perfectly, the idea behind catastrophic climate change is that CO_2 is an overwhelming *driver* of the global climate system and thus that its warming impact is predictable over time—in the same way that knowing the climate factors where I live, in Southern California, allows you to predict that it will be mostly dry, even though you can't predict exactly when it will rain.

Climate scientists who believe CO_2 is such a powerful driver feel confident in making *models*—simplifications—of the global climate system that predict its future based on CO_2 emissions.

Just about every prediction or prescription you hear about the issue of climate change is based on models. If a politician talks about "the social cost of carbon," that's based on model predictions. If an economist talks about "pricing fossil fuels' negative externalities," that's based on model predictions. If we hear dire forecasts of drought going forward, that's based on model predictions. Which means if the models fall, they are invalid. Therefore we need to ask the experts advising us an obvious and essential question: How good are the models at predicting warming or the changes in climate that are supposed to follow from warming?

One pitfall in asking this question is that we have to make sure we have evidence of models *predicting climate in advance*. Why do I say "in advance"? Because part of climate models involve "hindcasting" or "postdicting"—that is, coming up with a computer program that "predicts," after the fact, what happened. There are reasons to do this—namely, it's important to see if your model could have accounted for the past. *But a model is not valid until it makes real, forward predictions.* It's a truism in any field of math that if you are allowed enough complexity, you can engage in "curve fitting" for any pattern of data with an elaborate equation or program that will "postdict" exactly what happened in the past—but in no way does that mean it will predict the future. (Many investors lose money doing this sort of thing.)

The best way to test a model is to see whether it can make accurate and meaningful predictions about the future. In the last thirty years, the climate science community has had the opportunity to do that.

Many experts in modeling and in statistics thought this was an extremely dubious enterprise, given how complex the climate is—at least as complex as the economic system, where failed computer models helped promote policies that led to our recent Great Recession.

Consider perhaps the most famous model in the history of climate science, the 1988 model by James Hansen, who has a reputation in the media as the world's leading climate scientist. At twenty-four years old, the model has been given ample time to show its predictive accuracy. In the graph below, we can see how Hansen's prediction compares with the actual temperature measurements Hansen subsequently reported; he dramatically overpredicted warming.

Figure 4.2: The NASA/Hansen Climate Model Predictions vs. Reality

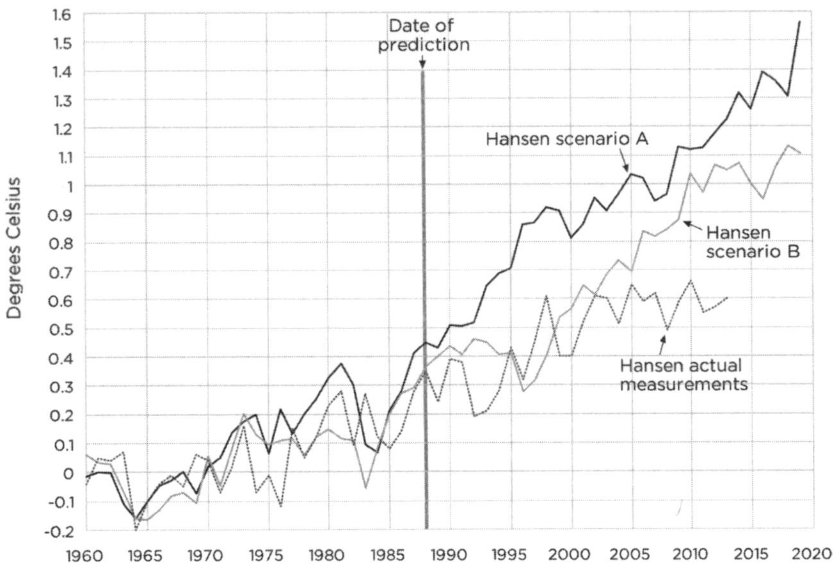

Sources: Hansen et al. (1988); RSS; Met Office Hadley Centre HadCRUT4 dataset; RSS Lower troposphere data

Note in particular that since the late 1990s, there has been no increase in average temperatures. Hansen and every other believer in catastrophic global warming expected that there would be, for the simple reason that we have used record, accelerating amounts of CO_2. But as the official government data show, these CO_2 increases have not driven major temperature increases; as CO_2 has increased dramatically, there have been relatively mild periods of warming, cooling, and now flattening. Thus, not just Hansen's model but every climate model based on CO_2 as a major climate driver has been a failure.

Here is a graph of 102 prominent, modern climate models put together by John Christy of the University of Alabama at Huntsville, who collects satellite measurements of temperature. Even though the modern models have the benefit of hindsight and "hindcasting," reality is so inconsistent with the theory that they can't come up with a plausible model. And note how radically different all the predictions are; this illustrates that the field of predicting climate is in its infancy.

Figure 4.3: Climate Prediction Models That Can't Predict Climate

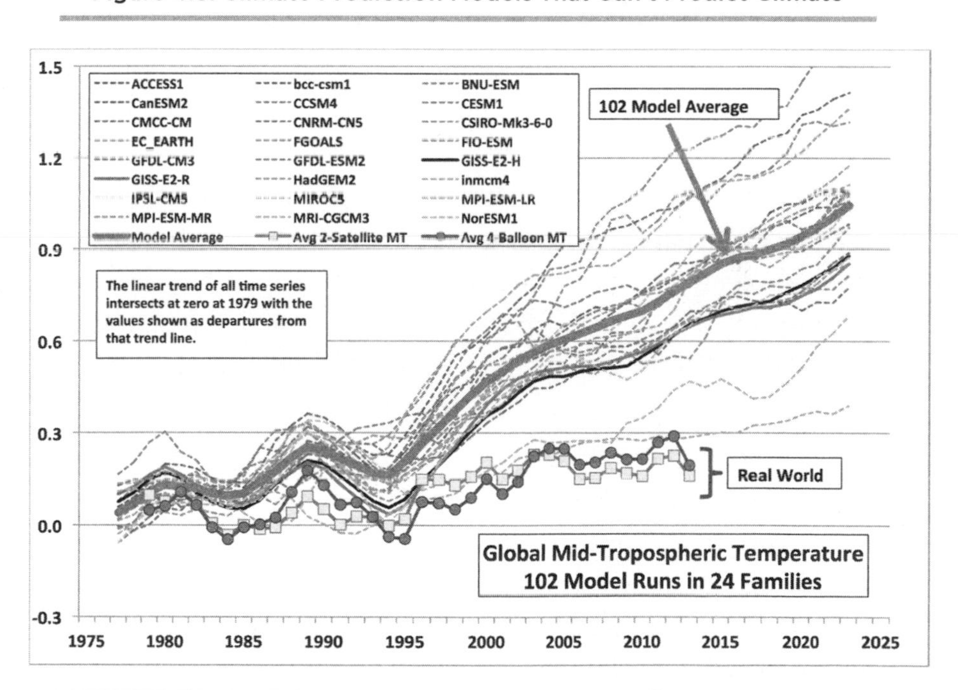

Source: Christy, Climate Model Output from KNMI, Climate Explorer (2014)

Here's the summary of what has actually happened—a summary that nearly every climate scientist would have to agree with. Since the industrial revolution, we've increased CO_2 in the atmosphere from .03 percent to .04 percent, and temperatures have gone up less than a degree Celsius, a rate of increase that has occurred at many points in history.[25] Few deny that during the last fifteen-plus years, the time of record and accelerating emissions, there has been little to no warming—and the models failed to predict that.[26] By contrast, if one assumed that CO_2 in the atmosphere had no major positive feedbacks, and just warmed the atmosphere in accordance with the greenhouse effect, this mild warming is pretty much what one would get.

Thus every prediction of drastic future consequences is based on *speculative models* that have failed to predict the climate trend so far and that *speculate a radically different trend than what has actually happened in the last thirty to eighty years of emitting substantial amounts of* CO_2. And we have not even explored the complete failure to make accurate predictions about climate changes in specific regions, which is what really matters in assessing and adapting to any climate-related threats.

If a climate prediction model can't predict climate, it is not a valid model—and predictions made on the basis of such a model are not scientific. Those whose models fail but still believe their core hypothesis right still need to acknowledge their failure. If they believe that their hypothesis is right and that complete lack of dramatic warming is just the calm before the storm, they should state all the evidence pro and con.

Unfortunately, many of the scientists, scientific bodies, and especially public intellectuals and media members have not been honest with the public about the failure of their predictions. Like all too many who are attached to a theory that ends up contradicting reality, they have tried to pretend that reality is different from what it is, to the point of extreme and extremely dangerous dishonesty.

CLIMATE DISHONESTY: EXTREME MISREPRESENTATION ABOUT EXTREME WEATHER

As predictions of extreme global warming have completely failed to materialize, there has been more of an emphasis on extreme weather as a reason to oppose fossil fuels. But this is misleading. The prediction of catastrophic climate change is based on the idea that *warming* will cause extreme weather.

And the data bear this out. As might be expected, given that there has been little warming, there has been little change in the trends of various types of storms. For example, here are the most up-to-date data as of mid-2014 on "Accumulated Cyclone Energy," which is what would need to increase if the frequency and/or intensity of storms were to increase. As the data show, this is, like most things in climate, a dynamic variable—one that shows no dramatic changes recently.

Figure 4.4: Storm Energy Is Normal

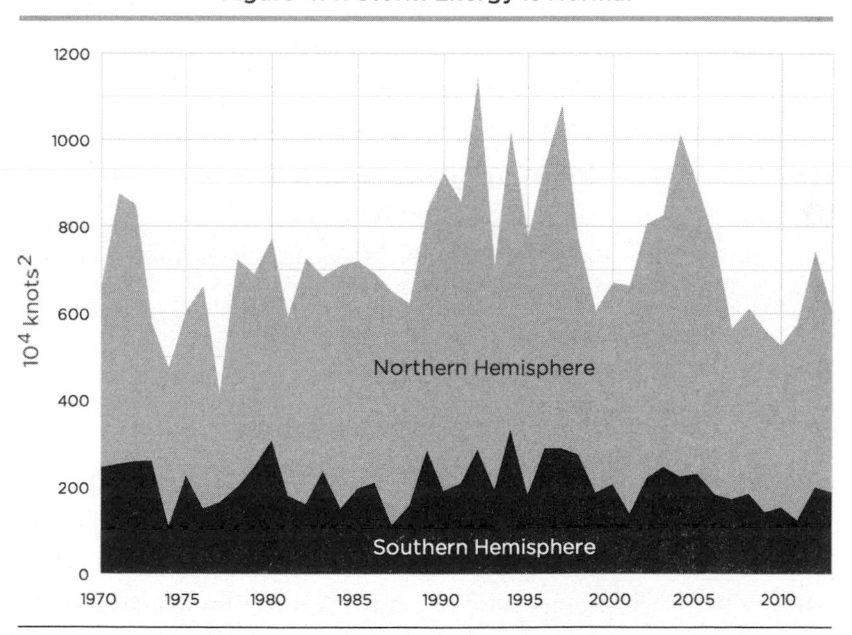

Source: Maue (2011, updated June 2014)

There is theoretical debate about how this would change if it had been warming dramatically—but it hasn't been warming dramatically.

Unfortunately, because people have been led to believe that CO_2 somehow causes climate change *in addition to,* not as a consequence of, global warming, it seems plausible to blame individual hurricanes on CO_2, even though the temperatures haven't increased. It is disingenuous for climate activists to blame every storm on climate change when there has been so little warming so far and when storm trends are so unremarkable. Remember, climate is always volatile, climate is always dangerous.

Or take the issue of sea levels, which we hear are rapidly rising. Al Gore's movie *An Inconvenient Truth* terrified many with claims of likely twenty-foot rises in sea levels.[27] Given the temperature trends, however, we wouldn't expect warming to have a dramatic effect on sea levels. And, in fact, it hasn't.

Figure 4.5 shows sea level trends from locations throughout the world. Note how smooth the trends are—and also notice how several of them are downward. This points to a truth about sea level and climate. It is affected by many factors, often factors that are much more important than any change in the global climate system.

But what about all those extreme scenarios of future sea level rise? They are not based on real trends or proven science; they are based on climate-prediction models that can't predict climate. And anyone who tries to equate science and speculation is being unethical. Which is, unfortunately, rampant.

CLIMATE DISHONESTY: EQUATING THE GREENHOUSE EFFECT WITH CATASTROPHIC CLIMATE CHANGE

The entire modern enterprise of catastrophic climate change predictions, the enterprise that threatens our energy supply, is based on equating a *demonstrated scientific truth*, the greenhouse effect, with extremely speculative projections made by invalidated models.

Figure 4.5: What Sea Level Rise Actually Looks Like

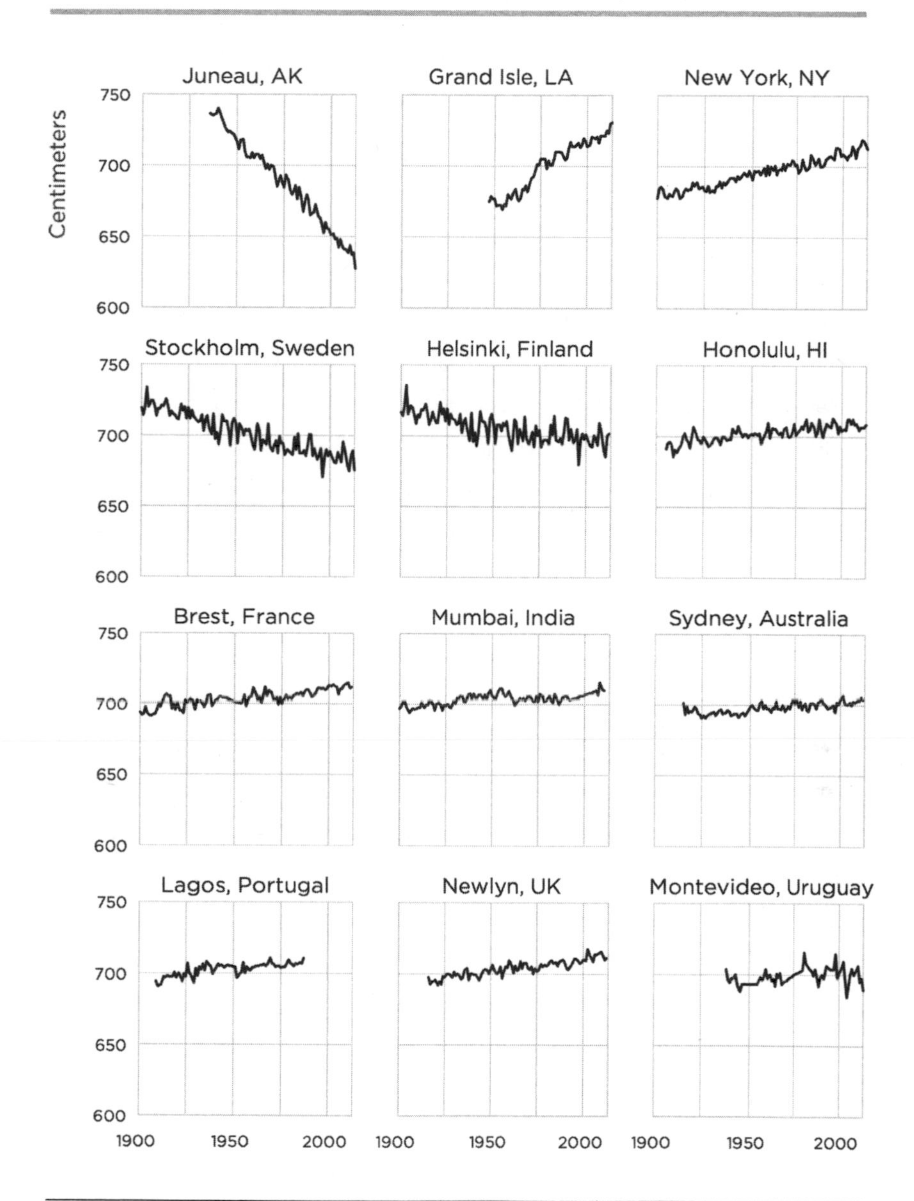

Source: Tide Gauge Data, Permanent Service for Mean Sea Level (2014)

In 1989, Bill McKibben pioneered this tactic in *The End of Nature*, wherein he called catastrophic climate change "the greenhouse effect." That would have been news to one of the discoverers of the greenhouse effect, Svante Arrhenius, who regarded increased CO_2 emissions as a very positive phenomenon. In 1896 he said: "By the influence of the increasing percentage of carbonic acid in the atmosphere, we may hope to enjoy ages with more equable and better climates, especially as regards the colder regions of the earth, ages when the earth will bring forth much more abundant crops than at present, for the benefit of rapidly propagating mankind."[28] (Remember this when we get to the fertilizer effect section below.)

Yet McKibben and others equate the greenhouse effect, dramatic global warming, and catastrophic global warming as it suits their political goals. By this kind of trickery, those who dispute catastrophic global warming are accused of denying the greenhouse effect and global warming. I experienced this in 2013 when I woke up to find myself named to *Rolling Stone*'s Top 10 list of "Global Warming's Denier Elite"[29]—in which they cited three articles of mine, each of which explained that CO_2 has a warming effect!

Here's what we know. There is a greenhouse effect. It's logarithmic. The temperature has increased very mildly and leveled off completely in recent years. The climate-prediction models are failures, especially models based on CO_2 as the major climate driver, reflecting a failed attempt to sufficiently comprehend and predict an enormously complex system.

But many professional organizations, scientists, and journalists have deliberately tried to manipulate us into equating the greenhouse effect with the predictions of invalid computer models based on their demonstrably faulty understanding of how CO_2 actually affects climate.

THE 97 PERCENT FABRICATION

This brings us to the oft-cited comment that 97 percent of climate scientists agree that there is global warming and that human beings are the main cause.[30]

First of all, this statement itself, even if it were true, is deliberately manipulative. The reason we care about recent global warming or climate change is not simply that human beings are allegedly the main cause. It's the allegation that man-made warming will be *extremely harmful to human life.* The 97 percent claim *says nothing whatsoever about magnitude or catastrophe.* If we're the main cause of the mild warming of the last century or so, that does not begin to resemble anything that would justify taking away our machine food.

But note how when I quoted John Kerry earlier, he went from "97 percent of climate scientists have confirmed that climate change is happening and that human activity is responsible" to "they agree that, if we continue to go down the same path that we are going down today, the world as we know it will change—and it will change dramatically for the worse."[31] Even in the 97 percent studies, which we'll look at in a moment, there is nothing resembling "97 percent of climate scientists have confirmed that . . . the world as we know it will change . . . dramatically for the worse." Kerry is pulling a bait and switch—using alleged agreement about a noncatastrophic prediction about climate to gain false authority for his catastrophic prediction about climate—and the anti–fossil fuel policies he wants to pass at home and abroad.

Unfortunately, this is very common. On his Twitter account, President Obama tweeted. "Ninety-seven percent of scientists agree: #climate change is real, man-made and dangerous."[32] There was no "dangerous" in the alleged agreement—and it wasn't "scientists," it was "climate scientists." This sloppy use of "science" as an authority, practiced by politicians of all parties, guarantees that we make bad, unscientific decisions.

On top of that, it turns out that the relatively mild "agreement" of the 97 percent is also a complete fabrication—which almost no one knows, because we're taught to obey authorities rather than have them advise us with clear explanations.

One of the main papers behind the 97 percent claim is authored by John Cook, who runs the popular Web site SkepticalScience.com, a virtual encyclopedia of arguments trying to defend predictions of catastrophic climate change from all challenges.

Here is Cook's summary of his paper: "Cook et al. (2013) found that over 97 percent [of papers he surveyed] endorsed the view that the Earth is warming up and human emissions of greenhouse gases are the main cause."[33]

This is a fairly clear statement—97 percent of the papers surveyed endorsed the view that man-made greenhouse gases were the main cause—*main* in common usage meaning more than 50 percent.

But even a quick scan of the paper reveals that this is not the case.

Cook is able to demonstrate only that a relative handful endorse "the view that the Earth is warming up and human emissions of greenhouse gases are the main cause." Cook calls this "explicit endorsement with quantification" (quantification meaning 50 percent or more). The problem is, only a small percentage of the papers fall into this category; Cook does not say what percentage, but when the study was publicly challenged by economist David Friedman, one observer calculated that only *1.6 percent* explicitly stated that man-made greenhouse gases caused at least 50 percent of global warming.[34]

Where did most of the 97 percent come from, then? Cook had created a category called "explicit endorsement without quantification"—that is, papers in which the author, by Cook's admission, did not say whether 1 percent or 50 percent or 100 percent of the warming was caused by man.[35] He had also created a category called "implicit endorsement," for papers that imply (but don't say) that there is some man-made global warming and don't quantify it.[36] In other words, he created two categories that he labeled as endorsing a view that they most certainly didn't.

The 97 percent claim is a deliberate misrepresentation designed to intimidate the public—and numerous scientists whose papers were classified by Cook protested:

- "Cook survey included 10 of my 122 eligible papers. 5/10 were rated incorrectly. 4/5 were rated as endorse rather than neutral."
  —Dr. Richard Tol[37]
- "That is not an accurate representation of my paper . . ."
  —Dr. Craig Idso[38]
- "Nope . . . it is not an accurate representation."
  —Dr. Nir Shaviv[39]
- "Cook et al. (2013) is based on a strawman argument . . ."
  —Dr. Nicola Scafetta[40]

Think about how many times you hear that 97 percent or some similar figure thrown around. It's based on crude manipulation propagated by people whose ideological agenda it serves. It is a license to intimidate.

CLIMATE ETHICS

The state of climate communication is a disgrace. Speaking from personal experience, it is incredibly difficult to get a straight answer about what is and isn't known in the field, because so much of it is catastrophic speculation by people who seem more focused on a political goal than on clear, honest, big-picture communication.

In 1996, Stanford climate scientist Stephen Schneider wrote an influential paper about the ethics of exaggerating the evidence for catastrophic climate change.

> On the one hand, as scientists we are ethically bound to the scientific method, in effect promising to tell the truth, the whole truth, and nothing but—which means that we must include all the doubts, the caveats, the ifs, ands, and buts. On the other hand, we are not just scientists but human beings as well. And like most people we'd like to see the world a better place, which in this context translates into our working to reduce the risk of potentially disastrous climate change. To do that we need to get some broad based support, to capture the public's imagination. That, of course, entails getting loads of media coverage. So we have to offer up scary scenarios, make simplified, dramatic statements, and make little mention of any doubts we might have. This "double ethical bind" we frequently find ourselves in cannot be solved by any formula. Each of us has to decide what the right balance is between being effective and being honest. I hope that means being both.[41]

I disagree entirely that this is a double ethical bind. It is doubly unethical. It requires deliberately misleading the public, which inevitably leads to uninformed, dangerous decision making.

We live in a society that has risen via the division of labor, by each of us specializing in, even mastering, some relatively small sliver of the ingredients of human survival and flourishing, so that in the aggregate we might create a world with an amazing sum of knowledge, technological achievement, and progress.

Specialization implies a sacred obligation. The specialist must never misrepresent what he knows and doesn't know, what he can do or can't do. The incompetent mechanic who claims that he can fix your complex engine problem, capitalizing on the fact that you know even less about engines than he does, is immoral.

In intellectual endeavors, in every field, there is an immense range of knowledge and opinion, from the decisively demonstrated to the wildly speculative. This is a good thing: Human knowledge builds on established knowledge, and each next step takes time to reach and establish. But specialists within the field have an obligation to explain precisely what they know and don't know—and also to welcome critical questioning and debate.

It can literally be deadly for a scientist to spread a hypothesis as fact. Take the realm of nutrition. For years, the government spread the gospel, treated as nutritionally proved, that a low-fat diet was healthy—a campaign that coincided with record obesity. I'm not going to claim that I know the perfect diet. The point is that, at this stage, no one appears to—and when scientists with speculative theories feel licensed to disseminate them as fact, it is the most irresponsible scientists who will often garner the most praise.

One such scientist is Paul Ehrlich, who writes: "Scientists need to be direct and succinct when dealing with the electronic media. One could talk for hours about the uncertainties associated with global warming. But a statement like 'Pumping greenhouse gases into the atmosphere could lead to large-scale food shortages' is entirely accurate scientifically and will catch the public's attention."[42] Is such a statement "entirely accurate scientifically"? What about the fact that were it not for the industry that necessarily emits greenhouse gases and were it not for the fact that Ehrlich's proposals to dismantle it were not followed, millions or billions would have died of starvation?

Imagine if we actually had a very serious problem from CO_2 emissions. These leaders would be doing everyone a disservice by exaggerating the evidence and leading many disillusioned followers to conclude that there was none.

If there was a true threat, they would win credibility by giving an honest, big-picture explanation to the effect of: We thought we didn't have to worry about CO_2 emissions from fossil fuels, but now we think there's strong evidence they could lead to something very dramatic and bad. Here's the evidence, and here are our answers to the counterarguments. We understand that there are other considerations, such as the crucial importance of fossil fuel energy in modern life, and we don't know enough about the big picture to say what policy should be, but we do think there's a major risk and we want to have a public discussion about it.

The lack of this kind of honest explanation and the conspicuous lack of concern about proposals to drain our energy suggest that they are not using human life as their standard of value. As does the fact that they do not publicize a significant positive impact of CO_2 emissions: global greening.

THE FERTILIZER EFFECT AND GLOBAL GREENING

Climate scientist Craig Idso is an anomaly. The son of climate scientist Sherwood Idso, he followed in his father's footsteps and did research on the most scientifically established—yet least discussed—aspect of CO_2's climate impact—the fertilizing effect of giving more CO_2 to plants.

Here's the situation from a plant perspective. Fossil fuels are super-concentrated ancient dead plants. When we burn/oxidize them, we increase the amount of CO_2, plant food, in the atmosphere. Thus, on top of getting energy, we should get a lot more plant growth—including growth of the most important plants to us, such as food crops.

Idso and others, conducting thousands of experiments in controlled conditions—where everything is held constant except CO_2—have convincingly demonstrated that more CO_2 means more plant growth.[43]

Figure 4.6 documents what happens to the four plants, identical and all grown at the same time, except with different levels of CO_2.

Again, the results are dramatic. If we are "green" in the sense of liking plant life, rather than in the sense of not affecting anything, shouldn't we be excited?

Figure 4.6: More CO_2, More Plant Growth

Photograph courtesy of Craig Idso, Center for the Study of Carbon Dioxide and Global Change

Worldwide increases in plant growth are nontrivial—indeed, Idso and others attribute significant portions of modern agricultural yields to increased atmospheric CO_2.[44] And there is a lot of evidence for this; observe the increases in crop yield when the following crucial crops were exposed to 300 ppm more CO_2 than is in the atmosphere.

Figure 4.7: More CO_2, More Crop Growth

Trees	Mean % Increase	Number of Studies
Black Cottonwood	124.0%	5
Red Maple	44.2%	13
Northern Red Oak	53.3%	7
Loblolly Pine	61.9%	65
Average	70.8%	

Fruits	Mean % Increase	Number of Studies
Cantaloupe	4.7%	3
Sweet Cherries	59.8%	8
Strawberries	42.8%	4
Tomatoes	31.9%	35
Average	32.8%	

Vegetables	Mean % Increase	Number of Studies
Green Beans	64.3%	17
Soy Beans	47.6%	162
White Potatoes	29.5%	33
Sweet Potatoes	33.7%	6
Corn	21.3%	20
Carrots	77.8%	5
Average	45.7%	

Grains	Mean % Increase	Number of Studies
Barley	41.5%	15
Rice	34.3%	137
Wheat	33.0%	214
Average	36.3%	

Source: Idso, Plant Growth Database (2014)

All of this follows from basic chemistry and biology. Below 120 to 150 ppm CO_2, most plants die, which means human beings would die. All things being equal, in terms of plant growth, agriculture, et cetera, more CO_2 is better. Today's climate gives us far less CO_2 than we would like from a plant-growth perspective. We would prefer the thousands of ppm CO_2 that, say, the Cretaceous period had.[45]

What's most important about all this is not that it proves that there will be overwhelmingly positive climatological effects from *increasing* CO_2—though I think that's a possibility. The climate system is complex, and if no one among the specialists can predict it well, I certainly can't.

What's most striking is that these extremely positive plant effects of CO_2 are *scientifically uncontroversial* yet *practically never mentioned,* even by the climate-science community. This is a dereliction of duty. It is our responsibility to look at the big picture, all positives and negatives, without prejudice. If they think the plant positives are outweighed, they can give their reasons. But to ignore the fertilizer effect and to fail to include it when discussing the impacts of CO_2 is dishonest. It is meant to advance an agenda by not muddying it with "inconvenient" facts.

Occasionally the fertilizer effect will be mentioned as a trivial impact, not worthy of discussion, because the greenhouse effect will allegedly outweigh it so much with "too much" heat. This is dubious, given the observable increase in plant growth under conditions of increased CO_2 and given that the heat predictions are failures.

What's also striking is how, even though we all know that plants live on CO_2, almost no one in the culture thinks of potential positive impacts when he thinks about his "carbon footprint." This is prejudice—the belief that man-made impacts on our environment are necessarily bad, that the standard of value is nonimpact, and that there's no possibility of improving on Mother Nature.

Given that the climate naturally changes and human beings have generally thrived the warmer it has been, it is quite possible that a higher global temperature with higher CO_2 levels would be a great boon. It makes no sense to believe that the unchanged climate is the ideal.

My reading of the evidence is that there is a mild greenhouse effect in the direction human beings have always wanted—warmer—and a significant fertilizer effect in the direction human beings have always wanted—more plant life. I believe that the public discussion is prejudiced by an assumption that human impacts are bad, which causes us to fear and disapprove of the idea of affecting climate, even though climate is an inherently changing phenomenon that has no naturally perfect state.

But with both of these, particularly the greenhouse effect, I think it's important to be open to new evidence and new developments. And the only way to do that properly is for the community discussing this, including the scientific community, to drop its prejudice against man-made impacts, stop thinking about being "effective," and think only about being honest.

5

THE ENERGY EFFECT AND CLIMATE MASTERY

FOSSIL FUELS' MOST IMPORTANT CLIMATE-RELATED EFFECT

So far we have surveyed the evidence about the impact of the greenhouse effect and the fertilizer effect on climate. Now we turn to a different question: How does the *energy* we get from fossil fuels affect the livability of our climate?

A theme of this book is that *energy is ability*—because energy can help us do anything better. So if we have more energy, all things being equal, we should be better at dealing with climate—at protecting ourselves from or counteracting storms, heat, cold, floods, and so on.

How much better? And how is that offset by risks? Well, we need a way of measuring climate livability.

One way to approach this is to look at overall life expectancy and income—the leading indicators of human flourishing. If our climate is a significant danger and has been getting more dangerous since catastrophic predictions began over thirty years ago, then its effect might show up; it certainly would if it had reached catastrophe status. But as we saw in chapter 1, the more fossil fuel we use, the more life expectancy and income we have.

But if climate danger was a growing threat that was at the earlier stages of a terrifying ascent, it wouldn't show up in life-expectancy statistics yet. Where it would surely show up is in statistics that measure climate danger specifically.

The best source I have found for this is called EM-DAT: The OFDA/CRED (U.S. Office of Foreign Disaster Assistance and Centre for Research on the Epidemiology of Disasters) International Disaster Database, based in Brussels.[1] It gathers data about disasters since 1900.

Here again is the graph from chapter 1 that relates CO_2 emissions, the alleged climate danger, to the number of climate-related deaths, which reflects the actual climate danger. It's striking: as CO_2 emissions rise, climate-related deaths plunge.

And to make matters better, in reality the trend is even more dramatically downward, because before the 1970s, many disasters went unreported. One big reason for this was lack of satellite data. Now we can see the whole world, so we can track icecaps and disaster areas with relative ease. In 1950, if there was a disaster in the middle of what is now Bangladesh, would that information have been accurately collected? In general, we can expect that in more recent years, more of the deaths were recorded and that in earlier years, fewer of the deaths were recorded. For some countries, there is simply no good data, because in underdeveloped places like Haiti or Ethiopia, we do not even know exactly how many people lived in a particular place before a disaster struck. Today we have much better information—and because disaster statistics are tied to aid, there is an incentive to overreport.

And the more we dig into the data, the stronger the correlations get.

Figure 5.1: More Fossil Fuels, Fewer Climate-Related Deaths

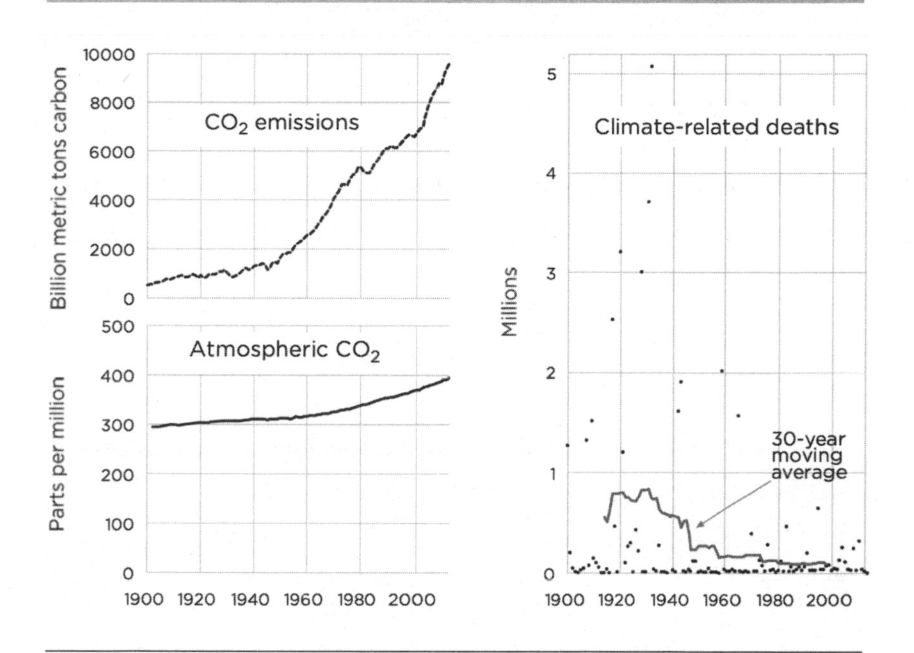

Sources: Boden, Marland, Andres (2013); Etheridge et al. (1998); Keeling et al. (2001); MacFarling Meure et al. (2006); Merged Ice-Core Record Data, Scripps Institution of Oceanography; EM-DAT International Disaster Database

Here are a couple of striking numbers from the data: In the decade from 2004 to 2013, worldwide climate-related deaths (including droughts, floods, extreme temperatures, wildfires, and storms) plummeted to a level 88.6 percent below that of the peak decade, 1930 to 1939.[2] The year 2013, with 29,404 reported deaths, had 99.4 percent fewer climate-related deaths than the historic record year of 1932, which had 5,073,283 reported deaths for the same category.[3]

That reduction occurred despite more complete reporting, an incentive by poor nations to declare greater damage to gain more aid, and a massively growing population, particularly in vulnerable places like coastal areas, in recent times.

All things being equal, one would expect the total number of deaths from these events to go up in proportion to population—

and if catastrophic climate change were true, we should see a massive recent uptick, not 29,404 deaths in 2013.

Just to be sure, let's look at the trends of individual types of climate danger in the last thirty years—when the predicted disasters were supposed to occur.

We'll start with droughts. Droughts are historically the most common form of climate-related death; a lack of rainfall can affect the supply of the two most basic essentials of life, food and water.[4]

Drought is supposed to be one of the most devastating consequences of CO_2 emissions, so let's see how they match up.

Figure 5.2: More Fossil Fuel Use, Fewer Drought-Related Deaths

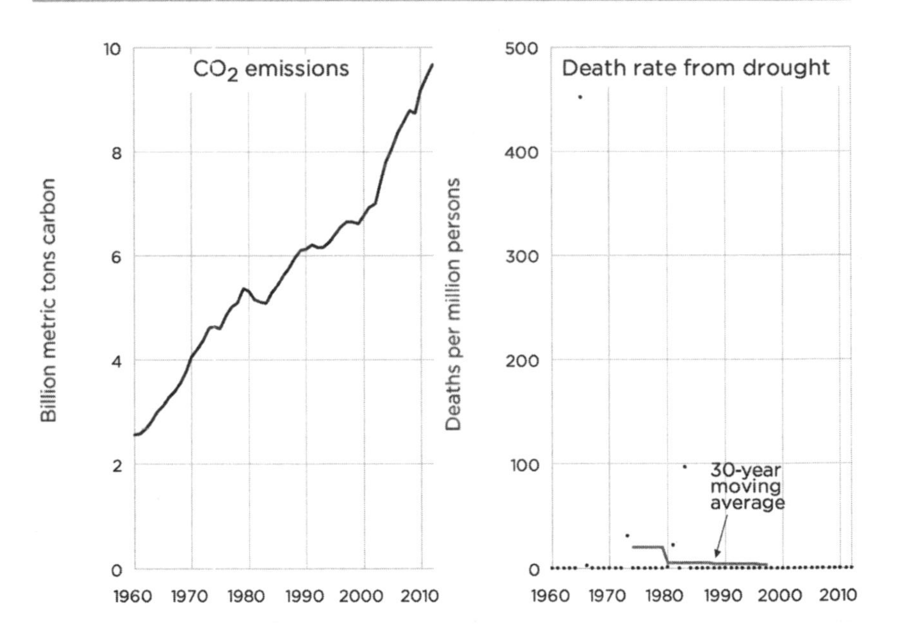

Sources: Boden, Marland, Andres (2010); EM-DAT International Disaster Database; World Bank, World Development Indicators (WDI) Online Data, April 2014

Clearly, CO_2 emissions have not had a significant negative effect on droughts, but expanded human ability to fight drought, powered by fossil fuels, has: from better agriculture (more crops for more people) to rapid transportation to drought-affected areas to modern irrigation that makes farmers less dependent on rainfall.

Shouldn't fossil fuel energy get some credit here?

What about dangerous storms?

Figure 5.3: More Fossil Fuel Use, Fewer Storm-Related Deaths

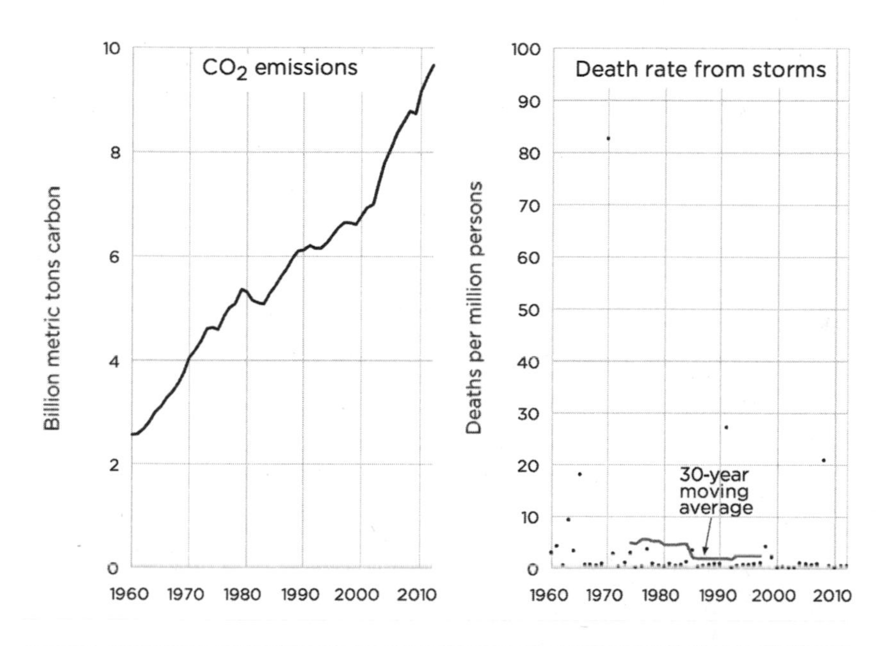

Sources: Boden, Marland, Andres (2010); EM-DAT International Disaster Database; World Bank, World Development Indicators (WDI) Online Data, April 2014

There are fewer storm-related deaths than ever. CO_2 emissions are not having a detectable effect on storm danger, but the fossil fuel energy that helps us build sturdy buildings and move people away from disaster areas, is.

What about floods?

Figure 5.4: More Fossil Fuel Use, Fewer Flood-Related Deaths

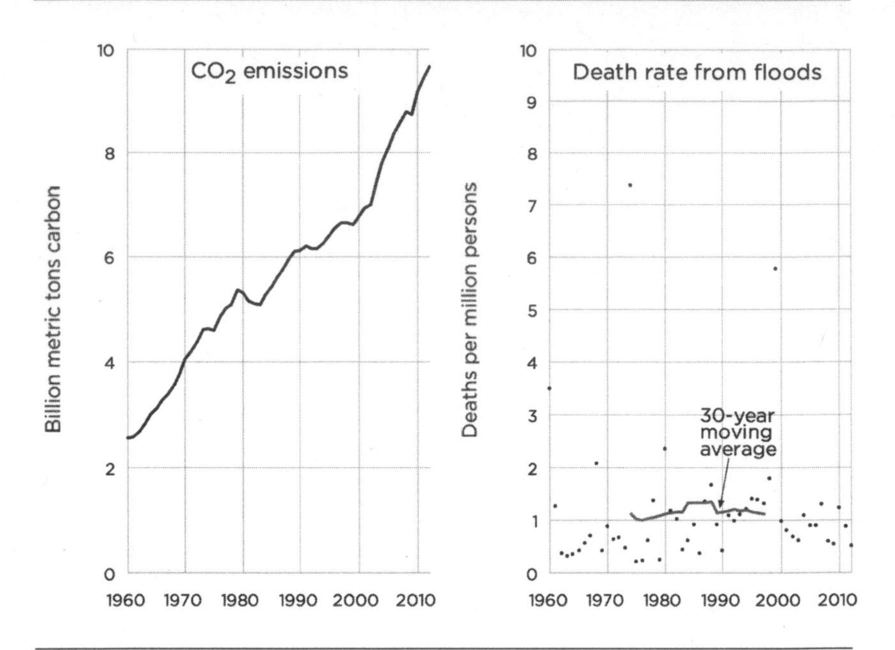

Sources: Boden, Marland, Andres (2010); EM-DAT International Disaster Database; World Bank, World Development Indicators (WDI) Online Data, April 2014

Again, the doomsayers are not looking at the big-picture data. If they were, they would appreciate the value of energy in building sturdier coastlines and better levees and seawalls. The more fossil fuel we use, the safer—dramatically, dramatically safer—we become from climate-related dangers.

We can also observe this from the perspective of comparing high-energy developed countries with low-energy underdeveloped countries. Here are the G7 countries compared to the world as a whole in death rates from climate-related causes.

When comparing storm death rates we see that developed nations fare much better than the world average. Note that the United States is making the G7 numbers higher because it is much more vulnerable to storms than the G7 average of countries. Of these countries, the United States is the only country with major tornadoes

Figure 5.5: More Development, Less Climate Danger

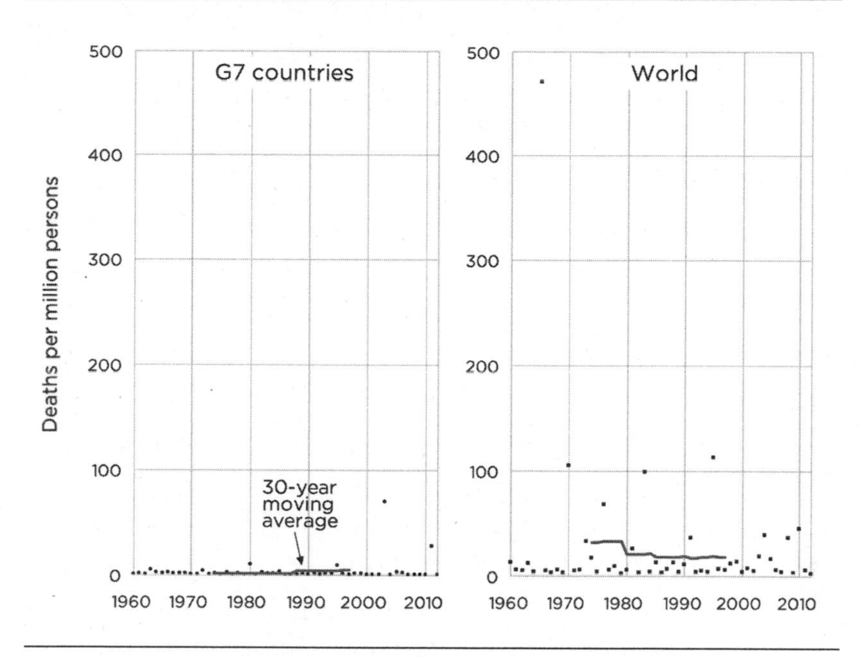

Sources: EM-DAT International Disaster Database; World Bank, World Development Indicators (WDI) Online Data, April 2014

and has received a significant number of landfalling hurricanes over the decades.[5]

If we look at year-to-year data, there is a dramatic difference between the heavy fossil fuel users and the light fossil fuel users in climate-related deaths—you are much, much safer in an industrialized country. This has a fairly obvious application: If you care about safety from climate, shouldn't you be encouraging rapid industrialization? Which today means encouraging fossil fuel use.

We can see that spikes from events like a major storm in one country increase the short-term volatility in the data. This reflects the nature of weather. The blue G7 nations are nevertheless much safer places, despite including disaster-prone nations with high population numbers, like the United States and Japan.

Shouldn't fossil fuel energy get some credit here?

To give you one particularly astonishing data point, the database reports that the United States has had zero deaths from drought in the last eight years. This doesn't mean there are actually zero, as the database only covers incidents involving ten or more deaths, but it means pretty near zero. Historically, drought is the *number-one climate-related cause of death.* Worldwide it has gone down by 99.98 percent in the last eighty years for many energy-related reasons: oil-powered drought-relief convoys, more food in general because of more prolific, fossil fuel–based agriculture, and irrigation systems.[6, 7] And yet we constantly hear reports that fossil fuels are making droughts *worse.* These reports give credibility to climate-prediction models that can't predict climate, but no credibility to the plain facts about how important more energy is to countering drought.

There is one more point to be made about the 29,404 deaths in 2013. Climate is no longer a major cause of death, thanks in large part to fossil fuels.[8] By contrast, there are 1.3 billion people with no electricity, the vast majority of whom will die early deaths, a problem that can be solved only by using more fossil fuels. Not only are we ignoring the big picture by making the fight against climate danger the fixation of our culture, we are "fighting" climate change by opposing the weapon that has made it dozens of times less dangerous.

The popular climate discussion has the issue backward. It looks at man as a destructive force for climate livability, one who makes the climate dangerous because we use fossil fuels.

In fact, the truth is the exact opposite; we don't take a safe climate and make it dangerous; we take a dangerous climate and make it safe. High-energy civilization, not climate, is the driver of climate livability. No matter what, climate will always be naturally hazardous—and the key question will always be whether we have the adaptability to handle it or, better yet, master it.

OUR NATURALLY HAZARDOUS CLIMATE

There is a widespread idea among climate commentators, including climate scientists, that the global climate system, absent human CO_2 emissions, is safe.

There is an unsophisticated and a sophisticated version of this argument.

Unsophisticated: John Kerry, when speaking to Indonesia, a nation that has dramatically increased its well-being in recent years through the burning of coal, tells them to stop burning coal: "But, ultimately, every nation on Earth has a responsibility to do its part if we have any hope of leaving our future generations the safe and healthy planet that they deserve."[9] But that "safe and healthy" planet is incredibly precarious for anyone outside high-energy civilization. Indonesia frequently gets hit by earthquakes and tsunamis killing hundreds or thousands and would be much safer were it more industrialized, with sturdy buildings, modern disaster relief, and the wealth and resources to rebuild quickly.

The sophisticated version of the idea that our climate is naturally safe or ideal says that because man has flourished in the current climatological period, the 10,000-year post–Ice Age stretch known as the Holocene, that is the only global climate we can live in and if there's a risk that fossil fuels will break the "natural" temperature highs of that last 10,000 years, we need to stop using them. "Just like us," says Bill McKibben, "our crops are adapted to the Holocene, the 11,000-year period of climatic stability we're now leaving . . . in the dust."[10]

This argument does not reflect reality.

First of all, the Holocene is an abstraction; it is not a "climate" anyone lived in; it is a *summary of a climate system that contains an incredible variety of climates that individuals lived in*. And in practice, we can live in pretty much any of them if we are industrialized and pretty much none of them if we aren't. The open secret of our relationship to climate is how good we are at living in different climates thanks to technology.

I live in the United States, in Southern California, which is naturally a near desert where I would have died of drought (or not lived here) in previous generations. But thanks to irrigation, air-conditioning, sturdy homes, and other technological advances (especially high-energy transport, which enables me to trade with people far away for goods I could not create under the local circumstances), this is one of the most wonderful places on Earth to live: I can enjoy warm, temperate, low-humidity weather without the downsides of the desert.

Southern California is considered a desirable climate to live in only because technology maximizes its benefits and minimizes its drawbacks. Technology enables us to live in practically any climate.

Consider that in the United States, a large country, we are home to every type of climate imaginable: from polar Alaska to desert California to swampy Florida to scorching Texas. And yet in each state we have a life expectancy of over seventy-five![11]

There is no climate that man is ideally adapted to, in the sense that it will guarantee him a decent quality of life. Nature does not want us to have a life expectancy of seventy-five or an infant mortality rate below 1 percent. Nature, the sum of all things on Earth, doesn't care about human beings one way or another and attacks us with bacteria-filled water, excessive heat, lack of rainfall, too much rainfall, powerful storms, decay, disease-carrying insects and other animals, and a large assortment of predators. Today we regard death before age thirty as a tragedy; in more "natural" times, it was the expectation.

We are naturally *dependent on* climate, and naturally *endangered by* climate.

While today it does not make sense to obsess about climate changes, at one point in history it did—because such changes controlled our lives more than we could control them.

The right climate conditions at the right time meant a good harvest—the wrong ones could lead to food shortages; the right climate conditions meant the ability to build at least a primitive civilization—the wrong ones could destroy that civilization in a few days.

To put it bluntly, in our "natural climate," absent technology, human beings are as sick as dogs and drop like flies. Notice that today, though we talk a lot about climate, and episodes of bad weather get huge media attention, we don't fear climate on a day-to-day basis.

There are two lessons here: First, weather, climate, and climate change matter—but not nearly as much as they used to, thanks to *technology*. Climate livability is not just a matter of the state of the global climate system, but also of the technology (or lack thereof) that we have available to deal with any given climate. Second, having that technology is useless unless we have the energy to run it.

We often talk about Mother Nature as if it is really our mother—a being that deliberately nurtures us and has our best interests at heart. But it isn't, and doesn't. Nature, including the climate, is a wondrous background that gives us the *potential* for an amazing life—if we transform it. To obsess over changes in the background while ignoring the need for technology and transformation is a prescription for a worse life.

The one thing we can't live without, climatologically, is technology. Which means we can't live without the fuel of technology, energy. Which means we can't live without energy we (and potentially everyone) can afford. Which means, for the foreseeable future—as in, most of the unrepeatable, irreplaceable years of our lives—we can't live without fossil fuel energy.

With it, we can achieve a stunning—and growing—amount of mastery over any climate hazards, natural or man-made. We have been doing so for decades. And we can get even better.

Let's look at how climate mastery applies to perhaps the most hypothetically dangerous consequence of hypothetically dramatic warming: significant sea level rises.

MASTERING THE SEA

The effect that seems to be most directly connected to the warming of the planet is the rise of sea level. If the planet warms enough ice from the polar regions will melt, adding additional water to the oceans, which will creep up the landmasses of the world, such as coastal cities and island nations—with the potential for massive population displacement and enormous numbers of "climate refugees."[12] As I indicated in the last chapter, all the actual evidence of the trends I have seen—as against model speculation—points to very mild sea level rises, much as we have had for thousands of years. But let's say sea levels rise more significantly—which has happened rapidly in the past due to natural causes. Some countries will face this situation no matter what; as we saw in Chapter 4, local conditions can lead to above average sea level rise (or fall) in certain regions—and land can sink independent of what the ocean does.

Fortunately, with development and cheap, plentiful, reliable energy, nations can transform their environments to be far safer.

Now, I'm not talking here about some science-fiction ten-foot-a-year rise in sea level, which would be an unmitigated disaster for coastal cities (the 2011 Japan tsunami paints a vivid picture of what this would look like). It would be a disaster because we would have had no time to plan. However, a rise of around two feet over a century, which the IPCC projects[13] (likely overprojects, given the models it relies on) is a much different proposition. People, even entire cities' worth, have time to find a solution.

History has provided us with an example of a people and nation that have experienced the problem of rising sea level: the Netherlands. More accurately, the Netherlands experienced a sinking of the land, rather than a rise of sea level, but the effect was essentially the same—50 percent of the Netherlands lies less than three feet above sea level, and roughly 20 percent of its people actually live at an elevation below sea level.[14]

This situation resulted from the choices that the residents of the Netherlands made about a thousand years ago. Early people drained the swamps of the region in an effort to find new farmland. While the peat soil beneath was very fertile, it was also very soft and began to sink as it was used for crops. In addition, peat (which is a precursor to coal) was also a very useful fuel, and the residents of the Netherlands extracted it and consumed it, causing the land to sink even more.[15]

At a certain point, it became obvious that floods were becoming a large problem because the water in these lowlands had nowhere to drain. Strangely (or perhaps not), this situation did not turn the residents of the Netherlands into helpless refugees but spurred them to find solutions.

As early as the year 1000, many villages in the Netherlands were connected by earthen walls that held the ocean and floodwaters back.[16] Over time, these primitive dikes were improved upon and, after several centuries, combined with pumps operated by windmills to remove any water that did manage to make its way into the lowlands.[17] These countermeasures kept the Netherlands mostly safe, except for rare large storms that overwhelmed the dikes and pumps.

Industrialization brought the flood control of the Netherlands, like every human endeavor, to a whole new level. Today the system consists of thousands of miles of dikes, dams, and electronically operated storm walls and sluice gates. Much of the system has been designed to withstand floods that have a probability of occurring once in ten thousand years. In addition, people are making plans and designs for what would have to change in the event of a rise in sea level.

Note that most of these innovations were made *before the availability of cheap, plentiful, reliable energy,* so if necessary, it would be far easier for other countries to replicate what the Netherlands has done. And who knows what else human ingenuity would come up with to deal with higher sea levels?

CLIMATE MASTERY

Even as we are taught to think of ourselves as in mortal climate danger, human beings are progressively becoming *masters* of climate. There are two components to mastering climate. One is control over the climate you're in. Two is the ability to make the most of the climate you're in.

At any point in the last billion years, the Earth has been full of all kinds of different climates with different levels and patterns of heat, cold, precipitation, etc.—and there will always be a wide variety of desirable and undesirable places. But even once human beings came on the scene, with their fantastic brains, they couldn't choose their climate very easily because of *lack of mobility*. Thanks to the internal combustion engine, which in 1992 Al Gore said should be outlawed in twenty-five years (i.e., 2017), we can go anywhere, anytime.[18]

You can also, incidentally, choose more dangerous areas that have other benefits. You can choose to expose yourself to hurricanes and flooding on the coasts because you like other features of the area. Or you can go to blizzard-prone areas because you want to ski and snowboard every day. This is the ultimate climate freedom. And we have this freedom, not just once but (to the extent we can afford it, which is closely related to the affordability of energy) throughout the year.

If you think of climate in a real way, not as some vague, mystical, "global climate" but as the climate around you, you are a *master of climate* just by virtue of the fact that you can change climates.

Of course, moving is not always easy (especially for the undeveloped world, which I'll discuss in a moment), but climate changes, even in the worst scenarios proposed by the most alarmist of the failed models, occur over periods of fifty to a hundred years. As with everything else in life, if we need to enhance our ability to do something—such as move—we need to be doubling down on energy production, not restricting it.

Again, mass movement with regard to climate changes seems very unlikely, but it's still worth mentioning because mobility is desirable, period, for the sake of the pursuit of happiness and because someday, some future generation *is* going to be faced with a dramatic climate change, and they'll need the energy and mobility to cope with it.

So fossil fuels give you the climate freedom to move, but as we have seen, they also give you the climate freedom to stay and thrive pretty much anywhere. Cheap, plentiful, reliable energy from fossil fuels amplifies our ability to build an infrastructure that insulates us from nature's climate dangers and discomforts. And cheap, plentiful, reliable energy from fossil fuels amplifies our ability to enjoy the benefits of a given climate (or multiple climates).

Bottom line: Fossil fuel energy, by enabling us to cheaply build and run wondrous machines that give us the mobility to choose any particular climate and the ability to increase the livability of that climate, has made us masters of climate. That doesn't mean we are invulnerable, but the numbers show that we have become progressively less vulnerable. And if we care about climate livability, energy and technology have to be the focus.

Why do our thought leaders never talk about this part of the fossil fuel–energy equation, which we can call the energy effect? It's all around us. While in Minnesota over New Year's 2014 visiting some dear friends (they would have to be dear for me to brave that weather), I realized, upon walking from my car to the bed-and-breakfast about forty feet away, that I couldn't find my key. I was in the natural climate. As I searched for my key at –10 degrees Fahrenheit, my fingers getting very cold very fast, it occurred to me that, were I stuck outside, I could easily die within the hour. I can only imagine what it would have been like the next week, when temperatures reached –70 degrees Fahrenheit one day. Fortunately, I could get warm in a high-energy car, find my key, and stay warm in a high-energy hotel.

Our climate focus needs to switch. The way to deal with climate danger is to take the high-energy actions necessary to deal with it. The answer is not in promoting inaction in the form of using *less* of our best form of energy. Once again, we have not been taught to think about these issues with human life as our standard of value.

BUT WHAT IF . . . ?

That said, I want to consider a hypothetical scenario in which CO_2 emissions *do* cause a significant climate danger around the world. I believe that even if that were true, the current conduct and policies of environmental leaders would be inappropriate.

Given what we know about the value of energy and of fossil fuels' superiority in most contexts, how should an honestly concerned person respond if there is a big problem?

First of all, by getting a straightforward understanding of the exact nature, magnitude, and certainty or uncertainty of the problem. It's actually hard to imagine a dilemma that might justify restricting fossil fuels, for our potential climate mastery is so great. But say there's a rapid rise in sea levels, enough to be truly concerned. What reaction would we want from our thought leaders?

One would be an embrace of technological solutions, including those used in the Netherlands and every other place that deals well with sea level and flooding. Another would be investing a huge amount of energy and technology looking for still better solutions.

In terms of communicating with the public, we would want our leaders to offer precise, objective briefings about evidence, risks, and probabilities with a recognition of the need to balance the risks with other risks (e.g., the hardships of energy loss). We would definitely not want vague talk of "catastrophe" with Hollywood hysteria scenarios.

We would want scientists and other thought leaders to welcome debate and be understanding of opponents. We would not want them to bash the inquisitive or skeptical as "deniers."

Economically, we would want a commitment to liberate any and every technology that could help, from seawall technology to dike technology to durable building technology to CO_2-free energy technology. We would not oppose the only globally scalable form of CO_2-free energy ever invented: nuclear power. I believe the evidence is clear that nuclear is the safest energy technology (safer than fossil fuels, hydro, wind, solar).[19] But even if it wasn't, if it would help avert a catastrophe, the doomsayers shouldn't be hostile to it. Ditto for large-scale hydroelectric power, which is also widely fought.

The one thing a human-focused response to a major climate danger would not do is try to save ourselves by pursuing solar, wind, and biofuels. These are the worst-performing sources of energy we have, and if we were truly in desperate straits, we would go with something that works; we wouldn't force everyone to use the worst and hope for the best.

Finally, on an emotional note, I think that a proper reaction to a major danger from fossil fuels would be *sorrow*. Think about it: If the energy that runs our civilization has a tragic flaw, that is a terribly sad thing. It would be even worse, say, than if wireless technology caused brain cancer. The appropriate attitude would be gratitude toward the fossil fuel companies for what they had done for us, combined with recognition that we would have to suffer a lot in the years ahead, combined with the commitment to the best technologies that I mentioned earlier.

But the doomsayers' response to the (fortunately) nonexistent tragedy is to express no gratitude for industrial civilization, and to condemn the fossil fuel companies as fundamentally evil. Bill McKibben calls them "Public Enemy Number One." James Hansen calls for them to be "tried for high crimes against humanity and nature." (Notice Hansen's equation of humanity and nature, making it unclear what his standard of value is.)[20] Others act is if it's an "exciting economic opportunity" to try to switch to the least competitive energy technologies on some insanely fast time frame, while opposing the truly effective energy technologies, such as nuclear, that could at least cushion the blow. "The winners of the race to reinvent energy will not only save the planet, but will also make megafortunes . . . fixing global warming won't be a drain on the economy. On the contrary, it will unleash one of the greatest floods of new wealth in history," says Fred Krupp, president of the anti–fossil fuel, antinuclear Environmental Defense Fund.[21]

Clearly the doomsayers are not really focused on minimizing CO_2 emissions. Clearly human life is not their operating standard of value; nonimpact is.

I believe that we owe the fossil fuel industry an apology. While the industry has been producing the energy to make our climate more livable, we have treated it as a villain. We owe it the kind of gratitude that we owe anyone who makes our lives much, much better.

There is one other issue of justice to discuss: the relationship between fossil fuel use and the climate difficulties of underdeveloped countries.

CLIMATE JUSTICE

One of the important moral issues of the climate-change discussion is the idea that the developed world is ruining the underdeveloped world by burning fossil fuels and that the solution is to stop using fossil fuels. This idea is usually accompanied by strongly emoted concern for the plight of the poor whose lives we are ruining.

If climate endangerment of the poor is a moral issue, then the climate catastrophists are major sinners.

We know that the way to make climate livable is not to try to refrain from affecting it but to use cheap energy to technologically master it. Thus, if the undeveloped world is having trouble dealing with climate, it's not because of our .01 percent change in the atmosphere; it's because they haven't followed the examples of China, India, and others who have increased fossil fuels use by hundreds of percent.[22] And the goal should be to help them do so—especially because the benefits of fossil fuels go far beyond climate: cheap, plentiful, reliable energy gives human beings the power to improve every aspect of life, including productivity, food, clothing, and shelter. You can't be a humanitarian and condemn the energy humanity needs.

Even if the underdeveloped world doesn't industrialize—which, by the standard of human life, it should—it is still wrong to claim that we're making lives worse climatewise (or otherwise). The data completely contradict that notion. Climate-related deaths are down 98 percent *worldwide,* including in undeveloped countries.[23] Our technologies and our wealth have given poorer countries better, cheaper everything: materials for building buildings, medicine, food for drought relief. The scientific and medical discoveries we have made in the time that has been bought with fossil fuel–powered labor-saving machines benefit everyone around the world.

To oppose fossil fuels is ultimately to oppose the underdeveloped world. Fortunately, many up-and-coming countries realize this. China and India and much of Southeast Asia are committing to technological progress, which means energy progress, which substantially means fossil fuel progress—and they don't appear to be willing to sacrifice their futures to climate fear. Neither should we.

THE BIG PICTURE

There is an incredibly positive story everyone should be told.

The climate future appears to be extremely bright. Fossil fuels' product, energy, has given us an unthinkable mastery over climate and thus record climate livability. And its major climate-affecting by-product, CO_2, has fertilized the atmosphere and likely brought some mild and beneficial warming along with it. But we can't know how good the warming is because, whether it is net negative or positive, it's completely drowned out by the net positive of the energy effect.

This will be challenged every day in the papers, by blaming storms on your tailpipe, by citing "studies" based on climate-prediction models that can't predict climate, but the truth is in the long-term trends and the powerful principles behind them.

The proper attitude toward human activity and climate is expressed in the 1957 novel *Atlas Shrugged* by Ayn Rand. Consider the following passage, where industrialist-philosopher Francisco d'Anconia remarks to steel magnate Hank Rearden how dangerous the climate is, absent massive industrial development. The conversation takes place indoors at an elegant party during a severe storm (in the era before all severe storms were blamed on fossil fuels).

> There was only a faint tinge of red left on the edge of the earth, just enough to outline the scraps of clouds ripped by the tortured battle of the storm in the sky. Dim shapes kept sweeping through space and vanishing, shapes which were branches, but looked as if they were the fury of the wind made visible.
>
> "It's a terrible night for any animal caught unprotected on that plain," said Francisco d'Anconia. "This is when one should appreciate the meaning of being a man."
>
> Rearden did not answer for a moment; then he said, as if in answer to himself, a tone of wonder in his voice, "Funny . . ."
>
> "What?"

"You told me what I was thinking just a while ago . . ."

"You were?"

". . . only I didn't have the words for it."

"Shall I tell you the rest of the words?"

"Go ahead."

"You stood here and watched the storm with the greatest pride one can ever feel—because you are able to have summer flowers and half-naked women in your house on a night like this, in demonstration of your victory over that storm. And if it weren't for you, most of those who are here would be left helpless at the mercy of that wind in the middle of some such plain."[24]

Think about the climate you live in. Think about how much the temperature changes every day and how uncomfortable or endangered you would be without climate control. Think of what even a garden-variety thunderstorm could have done to a farm or a home two hundred years ago—and then remind yourself that 1.3 billion people have no electricity today.

There is a group of people who are working every day to make sure that the machines that can make us safe from our naturally dangerous climate and enable us to thrive in it have all the energy they need. These people work in coal mines, on oil rigs, in laboratories, in boardrooms, all devoted to figuring out how to produce plentiful, reliable energy at prices you can afford—because that is what their well-being depends on and, in my experience, because they believe that it is the right thing to do. Those are the people in the fossil fuel industry, who are dehumanized in the media on a daily basis, who are tarred as Big Oil or, in the case of workers, such as coal miners, are portrayed as dupes who don't know what they're doing, who aren't wise enough to know they're making our climate unlivable through the work that supports themselves and their families.

Actually it is the top environmentalist intellectuals who lack climate wisdom. Because they are unwilling to think in an unbiased way about the benefits and risks of fossil fuels according to a human standard of value, they are blinded to the fact that the fossil fuel industry is the reason they're alive and not "helpless at the mercy of that wind in the middle of some such plain."

I wrote earlier that we owe the fossil fuel industry an apology for the way we've treated it on climate and that we owe them a long-overdue thank you. I meant it.

6

IMPROVING OUR ENVIRONMENT

ENVIRONMENTAL IMPROVEMENT

Try this thought experiment: Imagine that we transported someone from three hundred years ago, from essentially a fossil fuel–free environment, to today's world, which has fundamentally been shaped by coal, oil, and natural gas, and then took him on a tour of the modern world, good and bad, clean and dirty. What would he think about our environment overall?

I'll call our visitor Thomas, in honor of Thomas Newcomen, one of the pioneers of the steam engine, which was invented in 1712, almost exactly three hundred years ago.

Thomas's reaction would be disbelief that such a clean, healthy environment was possible.

"How is this possible?" he would ask. "The air is so clean. Where I come from, we're breathing in smoke all day from the fire we need to burn in our furnace."

"And the water. Everywhere I go, there's this water that tastes so good, and it's all safe to drink. On my farm, we get our water from a brook we share with animals, and my kids are always getting sick."

"And then the weather. I mean, the weather isn't that much different, but you're so much safer in it; you can move a knob and make it cool when it's hot and warm when it's cold."

"And you have to tell me, what happened to all the disease? Where I'm from, we have insects all over the place giving us disease—my neighbor's son died of malaria—and you don't seem to have any of that here. What's your secret?"

I'd tell him that the secret was his invention: a method of transforming a concentrated, stored, plentiful energy source into cheap, plentiful, reliable energy so we could *use machines to transform our hazardous natural environment into a far healthier human environment.*

Just as every region of the world, in its undeveloped state, is full of climate dangers (excessive cold, excessive heat, lack of rainfall, too much rainfall), so every region of the world is full of other environmental dangers to our health, such as disease-carrying insects, lack of waste-disposal technology, disease-carrying animals, disease-carrying crops, bacteria-filled water, earthquakes, and tsunamis. Nature doesn't even really give us clean air—because to live we have always needed some sort of fire, and for most of history, we had to breathe in smoke from outdoor fires or, once we got the benefit of true shelter, indoor fires, where the smoke was even worse, but the warmth was worth it.

To conquer these environmental hazards we need to *develop* a far more sanitary and durable environment. Development is the transformation of a nonhuman environment into a human-friendly environment using high-energy machines. Development means water-purification systems, irrigation, synthetic fertilizers and pesticides, genetically improved crops, dams, seawalls, heating, air-conditioning, sturdy homes, drained swamps, central power stations, vaccination, pharmaceuticals, and so on.

Of course, as I address in the next chapter, development and the fossil fuel energy that powers it carries risks and creates by-products, such as coal smog, that we need to understand and minimize, but these need to be viewed in the context of fossil fuels' overall benefits, including their environmental benefits. And it turns out that those benefits far, far outweigh the negatives—and technology is getting ever better at minimizing and neutralizing those risks.

How much of a positive difference does fossil fuel energy make to environmental quality? Let's look at modern trends in four key areas of environmental quality: water, disease, sanitation, and air.

Here's water quality—measured by the percentage of world population with "access to improved water sources."

Figure 6.1: More Fossil Fuels, More Clean Water

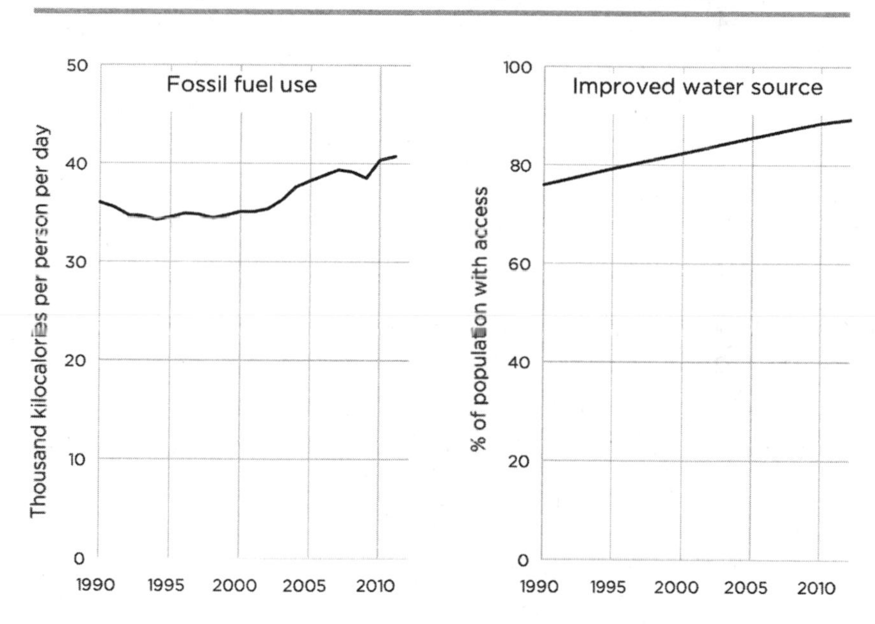

Sources: BP, Statistical Review of World Energy 2013, Historical data workbook; World Bank, World Development Indicators (WDI) Online Data, April 2014

Fossil fuel energy was essential to this improvement. It enabled us to transform once unusable water into usable water.

Most of Earth's surface is covered with water. The problem is that most of it is naturally in a chemical state unusable for our high standards and purposes. Most of the water is saltwater in the oceans. Most of the fresh water is trapped in massive ice sheets in places like Antarctica or Greenland. Some is part of a large water cycle of clouds and precipitation. Some portion is naturally "poisoned" brackish water of low quality in soil layers deep below the surface, containing too much salt and too many metals and other chemicals to be of any use without energy-intensive treatment. Nature does not deliberately or consistently produce "drinking water" able to meet a rigorous set of human health specifications.[1]

We need to transform naturally dangerous or unusable water into usable water—by moving usable water, purifying unusable water, or desalinating seawater. And that takes affordable energy.

If you were to turn on your faucet right now, in all likelihood you could fill a glass with water that you would have no fear of drinking. Consider how that water got to you: It traveled to your home through a complex network of plastic (oil) or copper pipes originating from a massive storage tank made of metal and plastic. Before it ever even got to the distribution tank, your water went through a massive, high-energy treatment plant where it was treated with complex synthetic chemicals to remove toxic substances like arsenic or lead or mercury. Before that, the water would have been disinfected using chlorine, ozone, or ultraviolet light to kill off any potentially harmful biological organisms. And to make all these steps work efficiently, the pH level of the water has to be adjusted, using chemicals like lime or sodium hydroxide.[2]

Natural water is rarely so usable. Most of the undeveloped world has to make do with natural water, and the results are horrifying. Billions of people have to get by using water that might contain high concentrations of heavy metals, dissolved hydrogen sulfide gas (which produces a rotten-egg smell), and countless numbers of waterborne pathogens that still claim millions of lives each year.[3] It's a major victory for any person who gains access to the kind of water we take for granted every day—a victory that fossil fuels deserve a major part of the credit for.

ERADICATING DISEASE

Potentially the worst, deadliest force in an environment is disease—the greatest predator of man. Some estimates have put the total number of human deaths caused by the bubonic plague, smallpox, and malaria alone at around one billion people.[4] While in the modern world we are taught to focus on any little particle emitted into the air by a power plant, we are not taught to appreciate the incomparably worse *diseases* those power plants have helped us get out of our air or made us safe from through mass production of pharmaceuticals and vaccinations.

Disease is on the decline—in large part because of the increased wealth that exists in the world and the increased *time* for scientific research—both products of cheap, plentiful, reliable energy. For example, Figure 6.2 illustrates the worldwide trend for tuberculosis, a major killer and one of the few diseases that is reported with any kind of consistency.

The tuberculosis trend just begins to indicate what is possible. Developed countries can use energy and technology to transform their environment to be totally rid of diseases that ravage underdeveloped countries today and that once ravaged all countries when they were underdeveloped.

While all infectious diseases can be traced back to some sort of pathogenic living organism, or germ, many of them require another

Figure 6.2: More Fossil Fuel Use, Less Tuberculosis

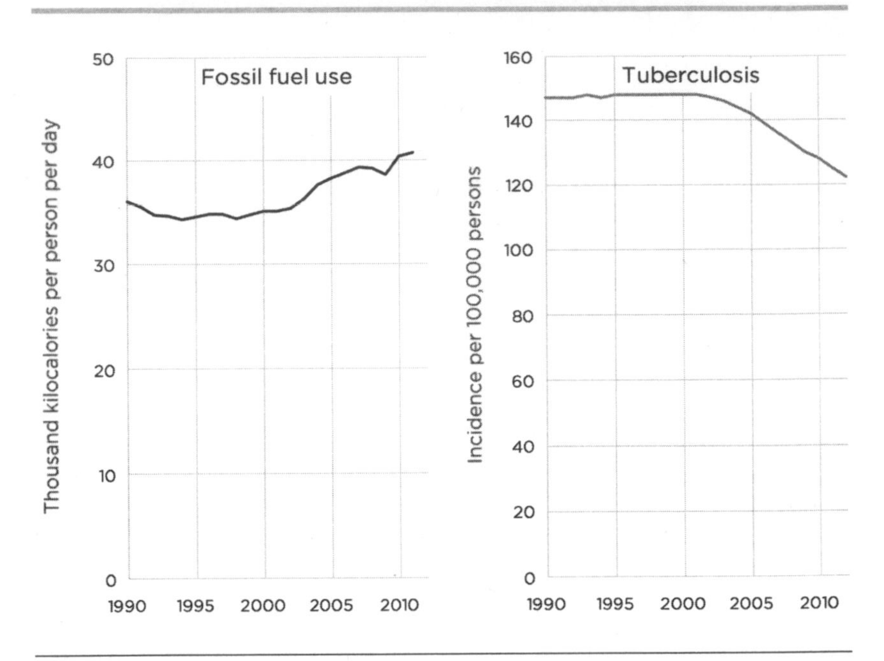

Sources: BP, Statistical Review of World Energy 2013, Historical data workbook; World Bank, World Development Indicators (WDI) Online Data, April 2014

animal to be transmitted: bugs. Mosquitoes transmit malaria and yellow fever, fleas transmit the bubonic plague, and lice transmit typhus. Once we understood this and had powerful machines to amplify our physical abilities *and our time to engage in scientific inquiry*, we declared war on the bugs that spread disease. We drained the wetlands where the bugs can lay their eggs, we introduced natural predators for the larval and adult forms of the bugs, we developed different chemicals that could kill the eggs, larvae, or adults, and we made it impossible for the bugs to encounter humans without encountering pesticide.

Notice how we discuss diseases like malaria as if they just happen to be in underdeveloped countries? Malaria existed in developed countries—they just developed their way out of it. Professor Paul

Reiter, a malaria expert who has publicly criticized the IPCC for blaming malaria on global warming, gave a memorable explanation of the history of malaria in front of the House of Lords:

> I wonder how many of your Lordships are aware of the historical significance of the Palace of Westminster? I refer to the history of malaria, not the evolution of government. Are you aware that the entire area now occupied by the Houses of Parliament was once a notoriously malarious swamp? And that until the beginning of the 20th century, "ague" (the original English word for malaria) was a cause of high morbidity and mortality in parts of the British Isles, particularly in tidal marshes such as those at Westminster? And that George Washington followed British Parliamentary precedent by also siting his government buildings in a malarious swamp! I mention this to dispel any misconception you may have that malaria is a "tropical" disease.[5]

Want an increasingly disease-free population around the globe? We need more cheap, reliable energy from fossil fuels.

SANITIZING OUR ENVIRONMENT

Historically, the inability to effectively deal with our own bodily waste has been one of the largest threats to human health. To this day it takes an enormous toll on human life throughout the world. For example, cholera is a bacterial disease that is transmitted through the ingestion of food or water contaminated by human fecal matter. The toxin that these bacteria produce inhibits the body's ability to absorb food and water, which can very quickly cause death through dehydration. Worldwide, over a hundred thousand people get sick from cholera annually. (Think about that when you hear environmentalists talk about "harmony with nature"—i.e., harmony with all our predators, their waste, and our waste.) But cholera has been all but eradicated in the industrialized world.[6]

Here's the big picture of sanitation—the percent of our world population with access to improved sanitation facilities, according to the World Bank.

Figure 6.3: More Fossil Fuel Use, More Access to Sanitation

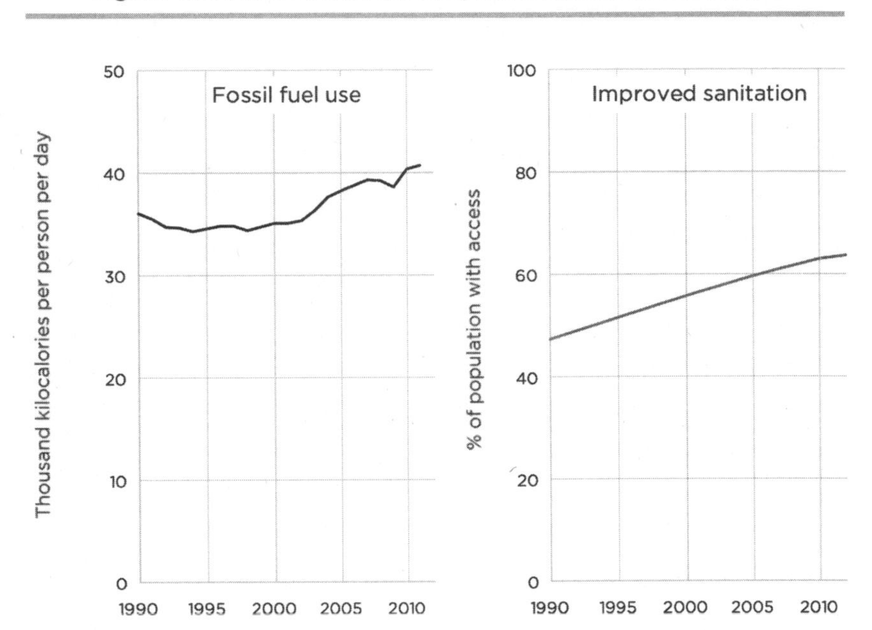

Sources: BP, Statistical Review of World Energy 2013, Historical data workbook; World Bank, World Development Indicators (WDI) Online Data, April 2014

Note that as recently as 1990, under half the world had "improved sanitation facilities." The increase to two thirds in only a few decades is a wonderful accomplishment, but a lot more development is necessary to make sure everyone has a decent, sanitary environment.

Part of the way we have solved sanitation problems is through the industrialized world's ability to thoroughly sanitize any water human beings might consume using high-energy machines. Just as important, we have created entirely separate water systems to deal with sewage. Historically, a person's sewer tended to be connected, at least in part, to his drinking water. This was rarely intentional, and early civilizations did construct sewer systems to isolate human waste, but natural, unrestricted water flows usually lead to a certain amount of mixing between the human waste and the nearest freshwater source—particularly as more and more people group together.

Today, sewage is not only kept separate from clean water sources, but it is also extensively treated to render its most dangerous elements harmless so that it can be disposed of safely, in some cases used as a fertilizer or even, thanks to the latest technology, turned into drinking water.[7] The technology of sewage treatment is another advance made possible by industrialization, and it is yet another energy-intensive process for transforming our environment.

Want a more sanitary environment for people around the globe? We need more cheap, reliable energy from fossil fuels.

CLEANING THE AIR

Most of us have had the experience of sitting around a campfire when the wind changes direction and blows the smoke into our faces right as we take a breath. The resulting experience is unpleasant: a few sharp coughs, along with some stinging of the eyes and throat. For us, it's a temporary annoyance. For billions of people around the world, it is an everyday experience.

Imagine if the only way you could avoid the danger of cold—historically, cold is a far bigger killer than warmth—was to light a fire in your house every day of the year. You could do things to reduce the amount of smoke you breathed in by using a chimney and opening windows (though at the expense of letting cold in), but the fact remains that you would be breathing in an enormous amount of smoke every day. For many people today, that's the choice: breathing in smoky air or cold.

Today the idea of using a fire to routinely heat our dwellings is foreign to most of us. Modern homes are heated with advanced furnaces that heat air within a machine and then send the warm air to various locations in the house. The heating is usually done either via clean-burning natural gas, in which case the furnace has an exhaust system to remove any waste from the combustion, or with electrical heating elements powered by mostly faraway smokestacks (which themselves minimize air pollution by diluting and dispersing particulates higher in the air).

The combination of sophisticated machines and cheap, reliable energy has made the heating of homes such a trivial issue that most of us have never considered its connection to cleaning up the air we breathe every day. And yet natural-gas furnaces enable us to enjoy all the benefits of having a warm place to live with none of the downsides of smoky, toxic air that our ancestors would have endured for the same privilege.

All of these benefits apply, not just in heating our homes, but in cooking our food. Indoor pollution from primitive cooking methods is a major global problem, and using fossil fuels can help solve it.

We need to consider all these air-cleaning benefits when we consider the air pollution *risks* of fossil fuels. Which is our next task.

7

REDUCING RISKS AND SIDE EFFECTS

THE POLLUTION CHALLENGE

Let's recap where we are. We use cheap, plentiful, reliable energy from fossil fuels to transform our environment to meet our needs. This leads to a far longer, more opportunity-filled life—and, it turns out, far greater safety from, even mastery of, climate. And the same holds true for environmental quality in general. We don't take a safe environment and make it dangerous; we take a dangerous environment and make it far safer.

But at the same time, we *do* create risks and side effects that can be deadly, and we need to understand them in order to set policies that will maximize benefits while minimizing risks. Like all technologies, fossil fuels have risks and side effects. When we transform those ancient dead plants into energy, bad things can happen.

Every time we use energy from fossil fuels (and from any other form of energy) we are engaging in a process that is filled with risk and that, if not managed properly, can become deadly. The process of producing energy can involve all manner of hazardous materials. For example, hydrofluoric acid, a vital material in certain kinds of oil drilling (and many kinds of mining) can literally travel through your skin and melt your bones.[1] The process of producing energy, because it involves something that can generate enormous amounts of power, always carries the risk of the power going out of control: explosions, electrocutions, fires. And then the process of producing fossil fuels involves by-products that can be hazardous to our health.

Take coal, the fossil fuel with the most potentially harmful by-products. Energy journalist Robert Bryce describes our "intense love-hate relationship" with "the black fuel."

> Coal heated people's homes and fueled the Industrial Revolution in England, but it also made parts of the country, particularly the smog-ruined cities, nearly uninhabitable. In 1812, in London, a combination of coal smoke and fog became so dense that according to one report, "for the greater part of the day it was impossible to read or write at a window without artificial light. Persons in the streets could scarcely be seen in the forenoon at two yards distance." Today, two hundred years later, some of the very same problems are plaguing China. In Datong, known as the "City of Coal," the air pollution on some winter days is so bad that "even during the daytime, people drive with their lights on."[2]

Stories of rampant smog in Chinese cities bring fears that the situation will inevitably get worse there and in any other country that industrializes. Fortunately, our experience in the United States illustrates that things can progressively get better.

Here again is a graph of the air pollution trends in the United States over the last half century. In the image are total emissions of what the EPA classifies as six major pollutants that can come from fossil fuels. Notice the dramatic downward trend in emissions—even though we were using more fossil fuel than ever.

Figure 7.1: Decline in U.S. Air Pollution

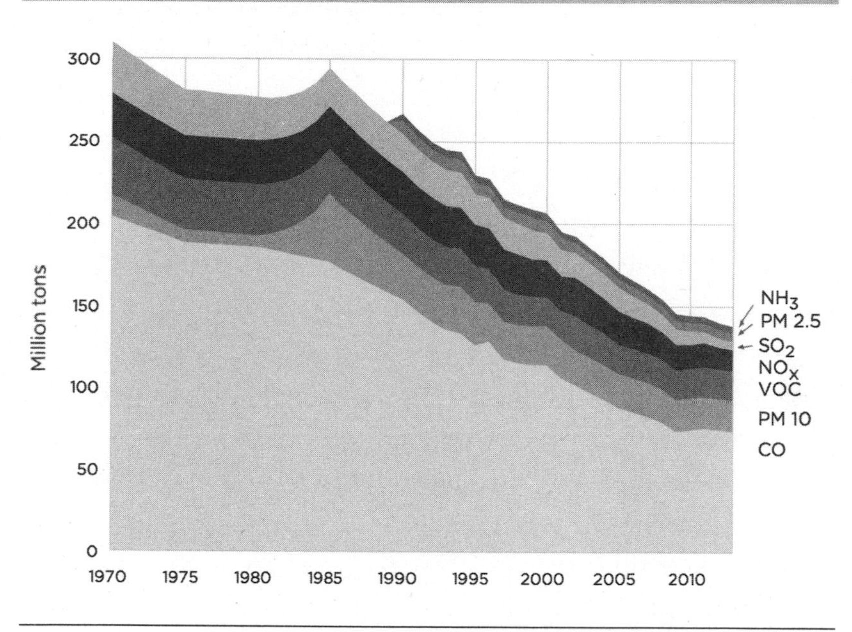

Source: U.S. EPA National Emissions Inventory Air Pollutant Emissions Trends Data

How was this achieved? Above all, by using antipollution *technology* to get as many of the positive effects of fossil fuels and as few of the negative effects as possible.

I like to think about risks and side effects this way. When we are using a technology, we are transforming our environment to meet our needs, to achieve a positive effect. But that transformation can accidentally or inevitably lead to an undesired effect—a power plant exploding or some type of molecule that, in high enough concentration, fouls up the air. The way to deal with it is to use technology to transform risks and by-products into smaller risks and smaller by-products.

To see how this works, let's take the fossil fuel that has historically and today been associated with the most environmental hazards: coal.

MANAGING BY-PRODUCTS AND RISKS—A UNIVERSAL CHALLENGE

Much of present-day energy discussion proceeds as if certain types of energy (fossil fuel, nuclear) are inherently dirty and dangerous and others (wind, solar) are clean and safe. But there is no limit to how much cleaner and safer fossil fuel use can be. For example, someday it might be possible to completely purify coal so that it generates no air pollutants and the materials that would have become air pollutants—nitrogen, sulfur, heavy metals—become valuable commodities. To a great extent, this is what we do with oil. What was once oil pollution dumped into a lake is now the basis for the plastic keyboard I am typing on.

At the same time, there is also no getting around the fact that *every* form of energy has risks—and every industry is responsible for managing them.

Consider the following story about the health and safety hazards of producing wind power. We think of wind as "clean" because there is no smoke coming out of the windmill. But in looking at any energy technology, we must remember that it's a process, starting with mining the materials necessary for the machines all the way to disposing of them. And wind turbines require far more toxic materials than fossil fuels do—materials called rare-earth elements. These elements are "rare," not in the sense that there are few of them, but in the sense that they exist in low concentrations in the Earth: it takes a lot of mining and a lot of separating of the desired metals from other elements using hazardous substances like hydrofluoric acid in order to get usable rare earth elements.

Here's what this process looks like in a major facility in China—where most rare earths for wind power are mined. This dispatch is from reporter Simon Parry, who visited to experience a rare earth mine firsthand. As you read his account, ask yourself: Does this mean that wind power is dirty and immoral?

> On the outskirts of one of China's most polluted cities, an old farmer stares despairingly out across an immense lake of bubbling toxic waste covered in black dust. He remembers it as fields of wheat and corn.
>
> ---
>
> Hidden out of sight behind smoke-shrouded factory complexes in the city of Baotou, and patrolled by platoons of security guards, lies a five-mile-wide "tailing" lake. It has killed farmland for miles around, made thousands of people ill and put one of China's key waterways in jeopardy.
>
> This vast, hissing cauldron of chemicals is the dumping ground for seven million tons a year of mined rare earth after it has been doused in acid and chemicals and processed through red-hot furnaces to extract its components.
>
> . . . When we finally break through the cordon and climb sand dunes to reach its brim, an apocalyptic sight greets us: a giant, secret toxic dump . . .
>
> The lake instantly assaults your senses. Stand on the black crust for just seconds and your eyes water and a powerful, acrid stench fills your lungs.
>
> For hours after our visit, my stomach lurched and my head throbbed. We were there for only one hour, but those who live in Mr. Yan's village of Dalahai, and other villages around, breathe in the same poison every day.
>
> ---
>
> People too began to suffer. Dalahai villagers say their teeth began to fall out, their hair turned white at unusually young ages,

and they suffered from severe skin and respiratory diseases. Children were born with soft bones and cancer rates rocketed.[3]

Does this mean the energy source this process makes possible is dirty and immoral? When I speak at colleges and students tell me that fossil fuels are "dirty," I sometimes ask them that question without first telling them what kind of energy the story is talking about. Inevitably they say it should be banned. When I reveal it's wind power, they protest, "No, just because something has problems doesn't mean we ban it. Otherwise we would ban everything. We should look at the big picture and try to solve the problem."

Exactly, I say. And we need to take the same approach with fossil fuels.

USING TECHNOLOGY TO MINIMIZE, NEUTRALIZE, AND REVERSE POLLUTION

Coal as a fuel has many advantages: Modern coal technology can harness coal energy extremely cheaply, and it is available in enormous quantities in many regions of the world. One of its disadvantages lies in its natural properties. As a solid fuel of condensed biological origin, it includes a lot of materials that were part of the natural environment long ago and that are potentially harmful to human health, such as sulfur, nitrogen, and heavy metals.

Fortunately, thanks to technology, coal has been getting healthier and cleaner since the 1800s, and today places that are home to coal plants, such as North Dakota, also have some of the world's cleanest air.

In the 1800s, coal was a major provider of energy for private households in Western countries, heating the stoves that were at the center of every home to cook and to keep the deadly cold outside. But it had a major direct health impact: the constant coal smoke indoors, which is almost always worse than any outdoor air pollution (although the coal stoves also led to plenty of outdoor air pollution). Urban areas were particularly affected, as they were the centers of industrial activity and at the same time densely populated. Pollution was visible as the smoke dampened the sunlight in the cities, darkened the laundry hanging to dry, and even blackened the trees with soot. Still, the energy from coal was so valuable that these side effects were more than tolerated. In many cases they were embraced.

Take Manchester, England, a major industrial city full of coal waste. There was no movement against air pollution in Manchester—even though its pollution makes China's air today seem pristine.

Why not? Because, as one commentator put it, the smoke was an "inevitable and innocuous accompaniment of the meritorious act of manufacturing."[4] No coal meant poverty and starvation—something to consider when we tell poor countries to adopt impractical technologies instead of coal.

In Manchester, the smoking chimneys were considered "the barometer of economic success and social progress."[5] That is not to say that living in smoke is the goal. Manchester did not want and would not need to live in coal smoke forever, and neither will the poor countries now striving for energy, thanks to enormous advances in coal plant technology.

In 1882 Thomas Edison revolutionized the use of coal, both from the production side and from the pollution side, when he built the first commercial, *centralized* coal-fired electric power plant for New York residents, starting what is the primary use for coal today: the production of electricity.

The power plant, being centralized, away from most people's homes, provided in-home power *without burning the coal in the house.* People went from burning coal in their home to getting power, almost by magic, from electricity that didn't pollute at all. (One negative side effect of centralized power is that we are not taught to think about what's "behind the plug"; many don't realize that it's "dirty" fossil fuels that make it possible for them to have clean electricity.)

Power plants also became progressively better at getting more energy from less coal, meaning less pollution (and lower energy costs).

One of the biggest problems coal can cause, particularly when paired with unfavorable weather conditions, is smog. As late as 1952, London experienced a massive air-pollution problem from a temperature inversion—a phenomenon that prevents particles from dissipating throughout the atmosphere and keeps them in dangerous, concentrated form. The particularly tragic 1952 inversion increased sulfur dioxide and soot concentrations all over the city, with a death toll estimated between four thousand and twelve thousand in a matter of weeks.[6]

Thankfully, technology has evolved greatly since then. Modern coal technology has many different means of reducing pollutants. There are filters like ceramics or fabric filtration systems to prevent undesirable substances from getting into the air; there are ingenious processes that use certain chemical agents, such as limestone, to bind pollutants and prevent them from escaping; there are mechanical devices like wet or dry scrubbers to separate out unwanted constituents; and there are many others.

Over time, these technological advances, as they became economical, became mandated by law. There is a whole controversial literature about which laws (federal or state) get credit, how much was industry's profit-motivated pursuit of efficiency, and to what extent the laws *overregulated* pollution at the expense of access to energy. For our purposes the important thing is this: It's clearly possible to increase fossil fuel use while decreasing pollution. And what applies to the most challenging fossil fuel, coal, also applies to oil and natural gas. This is a lesson that China can learn—and as it adopts more sophisticated modern coal plants, is starting to learn.

MINIMIZING DANGER

So far, we have discussed the challenge we face from the negative by-products that come from producing and using fossil fuels. But there is another category of risk: the danger of the energy itself going out of control.

On Deepwater Horizon—the oil rig that exploded in 2010, killing eleven workers and causing the BP oil spill—the energy went out of control.[7] In creating massive amounts of power, there's always the risk that we'll lose control of the power. This can mean a nuclear meltdown, a massive fire at an LNG terminal, an explosion in a coal mine, a downed live power line, or even a flying windmill.

When energy goes out of control, you can both lose the energy (sometimes permanently) and often lose lives. Obviously we want to avoid this as much as possible. Fortunately, modern technology has made energy production much, much safer. For comparison, in the 1870s, according to Daniel Yergin's *The Prize,* some *five thousand* people died annually in kerosene explosions from the lamps in their homes.[8] Gasoline is *more* volatile than kerosene, yet we drive our cars without any fear of explosion.

One consequence of the improvements in safety is far lower fatality rates for workers. According to the Bureau of Labor Statistics, someone in oil and gas extraction is one third as likely to get into a fatal accident as someone in logging and one fourth as likely as someone in fishing and hunting. And this might surprise you: The fatality rate in coal mining is, thanks to a concerted effort to radically improve safety over time, even lower than that of oil and gas extraction. If trends continue, both industries will become safer still over time.[9]

In 2013, having read some of my writing, the vice president of a coal company in Kentucky invited me to go underground in one of the mines. I eagerly accepted. When I got there, I was struck by how proud the workers were of their safety practices and how worried they were that *I* would be afraid to be underground. I reassured them that I was well aware of the statistics and that "the most dangerous part of a trip to a coal mine is the drive there." Statistically, that's absolutely right.

Every mining accident is a tragedy, but it should not be exploited to misrepresent the truth that coal is becoming safer and safer.

THE ROLE OF GOVERNMENT

The history of pollution laws consists of competing approaches to a challenging problem: how to protect the individual's right to be protected against pollution while simultaneously recognizing his right to pursue a modern, industrial life along with the energy that life requires.

My view of the right approach is: Respect individual rights, including property rights. You have a right to your person and property, including the air and water around you. Past a certain point, it is illegal for anyone to affect you or your property. But—and here's where things get tricky—it's not obvious what that point is. Let's look at two extremes.

One policy would be: People can pollute or endanger other individuals at will so long as they are viewed as benefiting "the common good." This policy, encouraged by some businesses in the nineteenth century, is immoral. It says that some individuals should be sacrificed for the business and its customers.

Here's another bad policy: *Any* amount of impact on air, water, and land should be illegal. This is simply impossible by the nature of reality—for example, consider that perhaps our most dangerous emission, contagious disease, can often be transmitted through the air or other life forms in ways we cannot detect or prevent.

At any given stage of development, some amount of potentially harmful waste cannot be prevented. For example, the man who invented fire could not protect himself or his neighbors from smoke. Should he be prohibited from using fire? Obviously not, because the *right* to be protected from pollution exists in a *context,* which is the right to the pursuit of life more broadly. Fire was far more helpful to human health than it was harmful, and so it was the right, healthy choice to use it.

Energy is so valuable that throughout history people have been willing to tolerate what we would consider intolerable pollution because the energy impact was so positive. Fortunately, over time, technology makes it possible to create less and less harmful waste and to better deal with the waste still created. As we have more wealth, energy, and knowledge, we can have stricter pollution standards and even more minimization of harmful waste.

The role of government is to pass laws based on individual rights and standards set according to science and the current state of technological evolution. The job of industry is to continue that evolution.

If the government does its job, it achieves two great results: the *liberation and growth of energy production* and the *progressive reduction of pollution and danger.* Historically, that is the trend—and with better laws and technologies here and abroad, we can do much better.

Unfortunately, we are taught the opposite.

It is a common practice to attack fossil fuels by misrepresenting them as fundamentally or uniquely dangerous. This is what's behind the current attack on fracking—hydraulic fracturing, part of the shale energy revolution I discussed in chapter 3.

There are at least four common fallacies used to discourage big-picture thinking and breed opposition to fossil fuels: the abuse-use fallacy, the false-attribution fallacy, the no-threshold fallacy, and the "artificial" fallacy.

These are things to be on the lookout for when you follow the cultural debate; they are everywhere, and all four are used to attack what might be the most important technology of our generation.

THE ABUSE-USE FALLACY

The largest fossil fuel controversy today, besides the broader climate change issue, is fracking—shorthand for hydraulic fracturing—one of several key technologies for getting oil and gas out of dense shale rock, resources that exist in enormous quantities but had previously been inaccessible at low cost.

Fracking has gotten attention, not primarily because of the productivity revolution it has created, but because of concerns about groundwater contamination. The leading source of this view is celebrity filmmaker Josh Fox's *Gasland* (so-called) documentaries on HBO.[10] Looking at how these movies have affected public opinion is an instructive exercise.

Both *Gasland* movies follow a similar three-part formula. First, Fox tells a sad story about a family undergoing a problem, usually with their drinking water. "When we turn on the tap, the water reeks of hydrocarbons and chemicals," says John Fenton of Pavillion, Wyoming. Then Fox blames it on the oil and gas industry's use of fracking—without exploring any alternative explanations, such as the fact that methane and other substances often naturally seep into groundwater. This is the false-attribution fallacy, which I'll discuss in a minute.

Even if Fox's examples were true, it would be illegitimate of him to conclude what he concludes today and what "fracktivists" demand—that fracking, and really all oil and gas drilling, should be illegal, as if any technology that can be misused should be outlawed.

Any technology can be abused. As we have seen, people are dying right now because of bad practices in the wind turbine production chain. It is irrational to say that because a technology or practice can be abused, it ought not be used.

I call this the abuse-use fallacy. It is a blueprint for opposing any technology. For example, Fox could make *Carland,* which could show car crashes and then blame all of them on "Big Auto." Then he could argue that because car crashes are possible, we don't need cars. In fact, Fox could make a far more alarming movie than *Gasland* based on supposedly risk-free solar and wind technology. Imagine a scene at a rare-earth mine in a movie called *Wasteland*.

Defenders of fracking often point out that the "abusers" Fox cities are false attributions—the next fallacy we'll discuss. But the pattern of argument would be wrong even if Fox wasn't fabricating particular abuses; individual abuses do not prove that an entire technology should not be used—they prove it should not be abused.

The abuse-use fallacy is deadly because it can be used to attack anything a group opposes. As citizens, we hate to see even one coal mine accident, one spill of hazardous liquids, one example of industry corruption, but we must use that feeling to advocate for proper laws and best practices, not to drive us to outlaw crucial technologies.

THE FALSE-ATTRIBUTION FALLACY

False attribution is claiming that one event causes another, devoid of proof. For example, in *Gasland*, Josh Fox famously showed people lighting their water on fire—a phenomenon that, unknown to many, is a frequent natural occurrence almost always stemming from the natural presence of methane (natural gas) in the water.[11] But it gets falsely attributed to fracking, as do many groundwater problems that are actually due to natural contamination of standard water wells.

A U.S. Geological Survey study conducted between 1991 and 2004 examined the quality of water from domestic wells and found: "More than one in five (23 percent) of the sampled wells contained one or more contaminants at a concentration greater than a human-health benchmark. . . . Contaminants most often found at concentrations greater than human-health benchmarks were inorganic chemicals, with all but nitrate derived primarily from natural sources."[12] In other words, more than one in five wells are naturally contaminated according to our government's standards. Yet we are taught to treat "natural" water as clean and blame all dirty water on industry, especially the fossil fuel industry.

Attributing water issues to fracking is almost always disingenuous. Here's the truth about groundwater. Every technology uses raw materials that must be mined from the ground, and anytime we drill or mine or dig underground, groundwater can be compromised. Of all the things you can do underground, fracking is the least likely to affect groundwater, because it takes place thousands of feet away from it. As President Obama's former EPA administrator Lisa Jackson acknowledged, there is no "proven case where the fracking process itself has affected water. . . ."[13]

If an oil company causes contamination at a fracked oil or gas well, it almost certainly has nothing to do with the fracking element of the process, but rather something near the groundwater, such as a surface spill of oil or some other liquid. So why single out fracking? Because attacks on fossil fuels thrive on technophobia—the fear of new technology—which is exploited by using unfamiliar, unknown terms like fracking. If Fox had opposed drilling, he wouldn't have gotten very far, because the public knows that, while accidents can happen while drilling, drilling itself is a vital human activity.

A more sophisticated version of false attribution uses prestigious studies based on speculative models. Just as climate discussions today are governed by speculative models whose (in)validity is rarely specified, so are pollution discussions. Regulators often use models that assert unprovable relationships between tiny amounts of particulate emissions and health problems.

The evidence is brought to us via "studies," cited by news media eager to run dramatic, "if it bleeds it leads" headlines. The main thing to watch out for here is a statement like "X causes Y"—e.g., "coal causes asthma." That's usually an oversimplification at best;

often it's completely bogus. It's hard to prove cause and effect. Here's a good question to ask when you encounter these kinds of claims: "Could you explain how you prove that—how you know that coal *in particular* caused asthma instead of everything else that might have caused it?" Usually the answer is no.

Let's look briefly at the claim that coal causes asthma problems through power plants' emission of particulate matter (PM).

Asthma or chronic respiratory disease has become more prevalent in Western countries.[14] That has triggered a variety of theories about the causes.[15] Claims about decreasing air quality or increasing exposure to toxins do not stand up, *as the increase in prevalence seems to be strongest in countries with much improved environmental quality;* for example, wealthier, cleaner West Germany had more asthma problems than poorer, dirtier East Germany.[16]

To put it in reverse, *countries with higher pollution levels have systematically shown lower rates of chronic respiratory diseases like asthma.* Something like asthma is a complex issue, and to use it to attack coal is to attack the health of everyone.

Or take mercury. Here's a summary of the typical argument about coal and mercury: Coal naturally contains mercury, a neurotoxin that can damage the nervous system, the brain, and other organs. When we burn coal, that mercury gets released into the atmosphere and ultimately rains down into bodies of water. This leads to higher mercury levels in fish, which lead to higher mercury levels in our bodies when we eat fish. Those levels are dangerous, particularly to the fetuses of pregnant women, whose children can experience developmental problems and learning disabilities. Therefore, coal is a massive threat to public health.

But here's the full context.

Mercury, a metal element, exists naturally throughout the world, most notably in the oceans, which contain an estimated 40 million to 200 million tons of mercury, as well as in most forms of plant and animal life. Mercury is released into the air by volcanoes, wildfires, and in far lesser quantities, the burning of coal. Natural causes of mercury are why the region of the United States with the highest mercury levels is the Southwest, whereas there are much lower levels in coal-heavy West Virginia and Kentucky.[17]

Mercury, like any substance, is toxic in certain forms and doses and harmless in others. The form of mercury that is of particular concern to human health is called methylmercury (or monomethylmercury), a combination of mercury, carbon, and hydrogen. Discussions of "mercury poisoning" are misleading, because mercury becomes methylmercury only under certain conditions, and methylmercury can be absorbed by human beings in relevant quantities only under certain conditions (for example, the element selenium seems to prevent the absorption of methylmercury).[18]

To be sure, negative cause-and-effect relationships do exist between fossil fuel emissions and human health—in certain concentrations and in certain contexts—but this doesn't appear to be one of them. Which brings us to the no-threshold fallacy.

THE NO-THRESHOLD FALLACY

All things are poison and nothing [is] without poison; only the dosage determines that something is not a poison.

—Paracelsus, sixteenth century[19]

The world around us and our own bodies consist of chemicals. All of them, without a single exception, can be poisonous to us if we are exposed to them in a certain concentration (which can be too high or too low) or in a certain form.

A simple example of this is medication. In the right concentration, a given hormone, heavy metal, or complex organic molecule can be lifesaving or can treat some nasty symptoms of a disease. Antibiotics are essentially poisons to microorganisms inside our bodies. If we take too much of certain drugs, we will die immediately.

The same is true for all substances in our bodies. Inside our bone tissue, for instance, there is a radioactive potassium isotope in a low concentration. Even pure water, which is the main constituent of our bodies, is a potential threat to our health. Drinking too much distilled water is dangerous because mineral-poor water entering our metabolism causes a mineral imbalance on the cellular level. On the other hand, pouring distilled, mineral-poor water on our skin is no threat.

A poison or pollutant is always a *combination* of substance and dose. If someone mentions just a substance to scare you, independent of the context or the dose, he has given you meaningless, misleading information. He is assuming or expecting you to assume that if a substance is dangerous in *some dosage,* it is dangerous in *all dosages.* One variant of this argument used to attack shale energy is the claim that fracking causes earthquakes.[20] This assertion is true in that fracking causes some amount of underground, earth-moving activity, but in almost all cases, it is completely inconsequential and not even discernible at the surface. A typical tremor that can be caused by hydraulic fracturing is –2 on the Richter scale, a "quake" that is not felt at the surface, causes no damage, and can be measured only deep underground. Such quakes are occurring continuously throughout the Earth, fracking or no fracking.[21]

What about a worst-case scenario? Many say that it's between 3 and 4 on the Richter scale, which means you can feel the quake but it's unlikely strong enough to cause damage.[22] And this is an incredibly unlikely scenario. For this we are supposed to ban all fracking?

Even if fracking in a certain place had a high risk of a truly dangerous earthquake—say, because it is near some seismically vulnerable area—that is an argument against fracking in that particular place, not an argument against fracking as such.[23]

When one treats something as poisonous regardless of dosage, he is denying the existence of a *threshold* at which a substance goes from being benign to harmful. If you deny a threshold, you can make a case for banning anything.

The no-threshold fallacy was used particularly insidiously in opposing nuclear power. People said we should have zero tolerance for radiation—not knowing, apparently, that the potassium in their bone tissue emits radiation, enough so that sleeping with a spouse gives you almost as much radiation as standing right outside a nuclear power plant. Both activities are nowhere near harmful.

"No-threshold" plus "false-attribution" is a dangerous combination in the hands of activists and regulators. They can keep claiming that nothing is clean enough and keep passing laws that regulate vital technologies, such as coal, out of existence. As always, whether we are talking about a natural substance or a man-made substance, our standard needs to be human life. *That* determines the threshold of danger.

THE "ARTIFICIAL" FALLACY

One of the big accusations against fracking is that it "uses chemicals." This is a funny way of putting it. Everything in our world uses chemicals, because our world is *made* of chemical elements.

The accusation is implying that fracking uses "artificial" or *man-made* chemicals, and the accusation assumes, and expects us to assume, that man-made means dangerous.

But it is simply untrue that "natural" is safe and man-made is unsafe. For example, fossil fuels are natural, organic, plant-based fuels whose pollution challenges stem from natural ingredients like sulfur and nitrogen and heavy metals. Arsenic and cyanide are natural substances, and many natural plants are poisonous.

The fact that we *didn't* make something shouldn't make us feel safe. And the fact that something is made in a laboratory shouldn't make us afraid. With every substance, we need to look at its nature and dosage in the context of human life.

One additional note: It *especially* doesn't make sense to be biased against man-made things, because they are *deliberately* made by a human mind, usually to promote human life. While man-made things can be bad, it is perverse to single out the man-made as bad per se. To be against the man-made as such is to have a bias against the *mind-made,* which is to be against the human mind, whose very purpose is to figure out how to transform our environment to meet our needs.

A HUMAN-CENTERED VIEW OF ENVIRONMENT

Fossil-fueled development is the greatest benefactor our environment has ever known. This needs to be mentioned in our environmental discussions, and so-called environmental groups need to be taken to task for omitting it. The only way fossil fuels are a net minus for "the environment" is if by "the environment" you mean our surroundings not from our perspective, but from a *nonhuman* perspective. From the perspective of organisms we need to kill or use to survive, such as the parasite, the malarial mosquito, the dangerous animal, or the trees we clear to build a road, *we* are a negative for the environment. (At the same time, we are positive for many other species. But as far as *our* environment goes, there is no environmental quality without development. And there is no global development without fossil fuel energy.

As we have seen, in using fossil fuels to improve our lives, including our environment, we create new environmental problems to solve—far fewer than those of undeveloped nature but real and important nonetheless. It's a fact of life that new technologies will bring problems that, by definition, would not exist if the technology didn't exist. There were no computer problems before computers. And just as we use computers to help solve computer problems, so we can use fossil fuels to help solve fossil fuel problems—to transform waste from a more dangerous form to a less dangerous form, or even to a benefit, by using energy and ingenuity. The energy we get from fossil fuels enables us to improve our environment—including mitigating or negating our own negative contributions.

This point belongs front and center in every discussion of fossil fuels and environment. All discussions of environmental issues need to recognize the phenomenon of *environmental improvement through development.*

Unfortunately, development has become one of the leading *targets* of environmental attacks. While any given instance of development can be bad—for example, if someone tramples on your property for his own project—the basic *purpose* of development is to improve our *human* environment.

That includes enjoying nature. Only with a society developed to the point of prosperity—including transportation systems crisscrossing the land, water, and air around the globe—can we enjoy the most beautiful parts of nature and the most fascinating parts of civilization.

The general opposition to development as antienvironment reflects a view that equates environment with *wilderness,* i.e., a nonhuman view of environment, which leads to an environment that is harmful for human beings because it does not sufficiently protect against natural threats or produce the resources necessary to overcome natural poverty. Here's the truth: The more development that happens, especially in underdeveloped countries, using fossil fuels, the more we can expect a *skyrocketing* of environmental quality around the world.

To be antipollution has nothing whatsoever to do with being antidevelopment. In fact, the two are incompatible; we need mass development to overcome nature's deadly pollution. And being prodevelopment, pro–fossil fuels, is completely consistent with another value that has been appropriated by the opponents of fossil fuels: appreciating nature.

PRESERVING NATURE TO BENEFIT HUMAN LIFE

In part because many anti–fossil fuel groups, such as the Sierra Club, celebrate the joys of spending time in less inhabited parts of nature, it is often believed that to advocate fossil fuel energy and the fossil fuel industry is to somehow oppose enjoying nature. (Terminology point: I consider human civilization just as natural as any other animal habitat, but I'll use *nature* in this context to mean "nonhuman nature.")

It's valuable to think of the ability to enjoy nature as a resource, something that we potentially have but don't automatically have. If we think that way, we see that like any resource, it is expanded by energy.

I'll use my own experience to illustrate. I have been fortunate enough to experience a wide variety of scenic, beautiful locations in my life. Some of my favorite moments are alone in nature. Snowboarding at the end of the day when I am the only one I can see on the mountain. Walking over lava on the Big Island of Hawaii. Standing under secluded waterfalls in the Grand Canyon. Like most people, sometimes I want to get away from everything, including all the complex machines.

And that's great, so long as I don't forget what got me there: complex machines.

Just as we are taught to think of nature as safe and clean, so we are taught to think of it as scenic. But it becomes scenic to us only if we have *access* to a variety of beautiful scenes.

There are more such scenes, but for most of history, no one got to enjoy many of them because they lacked the ultimate tool for enjoying nature—mobility. They also lacked the other crucial tool: adaptability, including medicine. Now we think of camping as a fun adventure. In the past, it was a deadly adventure.

If we view nature as another resource for us to enjoy and something to preserve when it is particularly beautiful or significant to us, then we will embrace fossil fuels. Fossil fuel energy gives us the *mobility* to get to it, the *adaptability* to be safe in it, and the *time* to enjoy it.

When we talk about resources, we have to remember that the only resource that can't be re-created, the real resource to guard jealously, is time—it is irreplaceable and unrepeatable. We can make more plastic, but we can't get back our time. And time is what enables us to enjoy nature. The more productive we are, the more time we have for leisure pursuits. (Whether people choose to take advantage of that is another issue.)

Furthermore, because fossil fuel energy is so dense and requires very little land and no live plants, it gives us both the wealth and the physical ability to preserve pretty much any piece of nature we want. And even in cases where one person's irreplaceable beauty is another person's needed energy source, we are talking about an installation that, if need be, has a finite lifetime and then can be transformed into a lush forest. Which is not to say that oil rigs are ugly—I think we should consider industrial civilization beautiful, too.

Look at the parts of the world where the "rain forest" (jungle) gets mowed down in seemingly shortsighted ways. Are they rich places? No, they are poor places with primitive agriculture and industry.

The now-developed world was once like that in preindustrial times. While we are taught to think that the country was once lush landscape, in fact, before coal, oil, and natural energy, our country and others survived by developing the landscape. As geographer Pierre Desrochers writes:

> Carbon fuels made this expansion of the forest cover possible in various ways. With the development of more sophisticated combustion technologies, coal, heavy oil and natural gas proved vastly superior alternatives to firewood and charcoal. Through their role as long-distance land and maritime transportation fuels, coal and later petroleum-based fuels (diesel and marine

bunker fuel) encouraged agricultural specialization in the most productive zones of the planet, in the process making much marginal agricultural land superfluous.[24]

If you love enjoying nature, you should love fossil fuels.

The same basic logic applies to more abstract concerns about "biodiversity" and species extinction. There are huge debates in the ecology literature about what is happening or not happening to what species, and I have not studied them enough to take sides, but I can say that from an energy perspective, to the extent it makes sense to preserve a given species or biological arrangement—and such decisions should be made according to a human standard of value, not a nonhuman one—cheap, plentiful, reliable energy gives us the means to do so just as we can preserve a desirable forest or park. It is only when we are truly living off the rest of nature that we must gobble up whatever we can.

Whether to actively preserve a species or not should be made with reference to a human standard of value. Much of the ecology field holds to the nonimpact standard, which treats another species' extinction as intrinsically wrong. But human beings are right to favor some species over others. For example, pigs, cattle, and chickens are in no danger of extinction because their human-centered benefits are immediately visible, so we make them some of the most abundant life-forms on Earth. On the other hand, wolves and bears and disease-carrying insects have been threats that we destroyed in many regions. There is no inherent reason to think that the extinction of any given plant or animal is bad for humans. We should focus on maximizing our benefits. That can be the removal of a direct threat, such as making bears nonexistent where our kids go to school, or the preservation of species that we want to survive, such as the panda, even if we do not strictly need it for our own survival.

THE BIG PICTURE

So far we have seen that the overall impact of fossil fuels on our environment is tremendously positive. But let's step back and ask this: Why are we concerned about our environment? Why are we concerned, say, about pollution? Of course, most fundamentally we desire human flourishing but in particular we desire human *health*. Therefore, in looking at fossil fuels and environmental quality, it's important to look at not just how they help us transform our environment for the better, but also how they help us transform *ourselves* for the better through health technology.

Let's look at the trends: infant mortality, mortality under five, malnutrition, and life expectancy.

Figure 7.2: Health Trends Improving Across the Board and Around the World

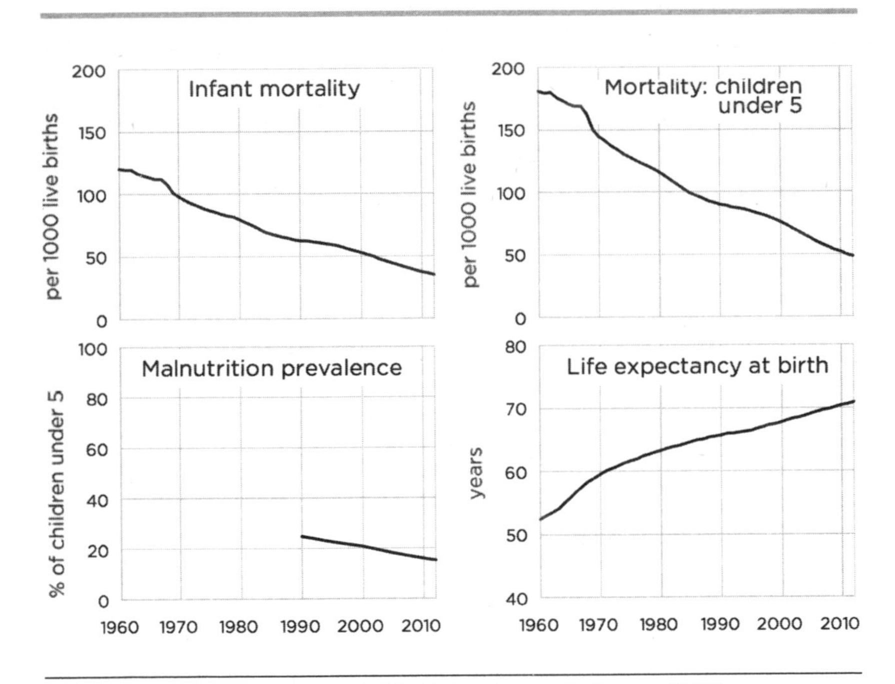

Source: World Bank, World Development Indicators (WDI) Online Data, April 2014

Every one of these graphs represents a collection of real people, many of whom have been recently empowered by energy and many of whom who are suffering every day for lack of it.

World life expectancy at birth has gone up from sixty-three in 1980 to seventy in 2012. The child mortality rate on the planet went down from 115 to 47 per 1,000 live births. Infant mortality declined from 80 to 35 per 1,000 live births in the same time period.[25] The incidence of tuberculosis, an infectious disease that particularly threatens poor people with little access to modern medicine, has declined from 147 per 100,000 population in 1990, when the World Bank's record begins, to 122 in 2012.[26] Malnutrition, defined by the percentage of children under five with significantly below average weight or height for their age, has been constantly decreasing at a significant rate since 1990.[27] Access to electricity and improved water sources, which are basic indicators for human well-being, hygiene, and health in general, went up as well.[28]

Developing countries in the sub-Saharan and East Asian region have been particularly impressive; East Asian developing countries now have an average life expectancy at birth of seventy-three years. There is much credit to be given to industrial-scale energy, primarily, as we have seen in previous chapters, from fossil fuels. Without a large amount of affordable energy, the vast majority of the people whose lives were drastically improved in recent decades would still sit in the dark mourning their dead children and friends, if they were ever born in the first place.

Many energy-intensive technologies influence our overall health in a positive way. Food production, modern medicine, and sanitation require cheap, plentiful, reliable energy to make them available and affordable to as many people as possible.

All of this is part of the big picture of fossil fuels' impact on our lives, health, and environment.

To summarize, fossil fuels improve our environment by, among other things, empowering us to fight the otherwise overwhelming health hazards of nature. Like all forms of energy, they have risks and by-products, but they also give us the energy and resources to minimize, neutralize, or even reverse those harms. More broadly, if health is our concern, fossil fuels underlie the food and medical care systems that have created the longest life expectancy in history.

Once again, we see that an alleged negative of fossil fuels, its impact on environmental quality, is in fact a tremendous positive.

We have a choice to make. Will we use fossil fuels to maximize human well-being in all areas of life, including our environment? Or will we continue to see fossil fuels only through negative glasses, blind to the tremendous benefits that have come so far, and the tremendous ones that can come in the future?

8

FOSSIL FUELS, SUSTAINABILITY, AND THE FUTURE

IS OUR WAY OF LIFE SUSTAINABLE?

Exploring the evidence about mankind's use of fossil fuels so far, we have seen that the fossil fuel industry is far and away the world leader at producing cheap, plentiful, reliable energy and that that energy has radically increased our ability to create a flourishing society, a more livable climate, and greater environmental quality. On these fronts, so long as we are able to use fossil fuels, the evidence is overwhelming that life can get better and better across the board, as we use fossil fuel technology and other technologies to solve more problems—including those that fossil fuel technology and other technologies create.

One big question remains: What are the *long-term* prospects for this way of life? While today we are rich in fossil fuel resources and the wealth they help us create, what is in store for the future?

With so much consuming, can this way of life really last? Is it sustainable?

The answer is better than yes. Not only can our way of life last; it can keep getting better and better, as long as we don't adopt "sustainability" policies.

In chapter 3, we saw that the amount of unused fossil fuel raw material currently in the Earth exceeds by far the amount we've used in the entire history of civilization by many multiples and that the key issue is whether we have the technological ability and economic reason to turn that raw material into a resource.

For years, actually centuries, opponents of fossil fuels—and some supporters of fossil fuels—have said that using fossil fuels is unsustainable because we'll run out of them.

Instead, we keep *running into them*. The more we use, the more we create. Fossil fuel energy resources, we discussed, are *created*—by turning a nonresource raw material into a resource using human ingenuity. And there is plenty of raw material left.

In the last few years, the shale energy revolution has unlocked vast new oil and gas resources, making the "running out of fossil fuels" claim seem implausible for the foreseeable future. Many environmental leaders have therefore shifted from saying that we're running out of fossil fuels to saying that our abundance of fossil fuels is causing us to run out of *other* resources—arable land and water, most alarmingly, but also a whole host of other materials that are crucial for civilizations.

"Consuming three planets' worth of resources when in fact we have one is the environmental equivalent of childhood obesity—eating until you make yourself sick," says David Miliband, secretary of state for the environment, food, and rural affairs in the United Kingdom.[1] In response to criticisms of renewable energy plans as utopian and far-fetched, Bill McKibben says, "Perhaps it's the current scheme, with its requirement of endless growth in a finite world, that seems utopian and far-fetched."[2]

The theory behind these predictions is that Earth has a finite "carrying capacity," an idea that was spread far and wide in the 1970s. Two of the leading exponents of this view were Paul Ehrlich and John Holdren. In their landmark book, *Global Ecology,* they wrote:

> When a population of organisms grows in a finite environment, sooner or later it will encounter a resource limit. This phenomenon, described by ecologists as reaching the "carrying capacity" of the environment, applies to bacteria on a culture dish, to fruit flies in a jar of agar, and to buffalo on a prairie. It must also apply to man on this finite planet.[3]

These theories were not idle banter—they were used by many to call for drastic restrictions on fossil fuel use, much as we have today. Ehrlich and Holdren announced, "A massive campaign must be launched to restore a high-quality environment in North America and to de-develop the United States."[4] This meant an attempt to reverse industrial development—by law: "This effort must be largely political."[5]

These ideas were viewed highly enough that Holdren's body of work, which stresses these themes over and over, gave him the prestige to become science adviser to President Barack Obama.

As we've discussed earlier, these predictions were wrong, but why, exactly, were they wrong? The most direct reason is that there are far more fossil fuel raw materials and far more human ingenuity to get them than Ehrlich and Holdren expected. But there is a deeper error here, an error at the root of the whole concept of sustainability. The error is a backward understanding of resources.

THE UNLIMITED POTENTIAL FOR RESOURCE CREATION AND HUMAN PROGRESS

The believers in a finite carrying capacity think of the Earth as something that "carries" us by dispensing a certain amount of resources. But if this was true, then why did the caveman have so few resources?

Those who believe in the ideal of human nonimpact tend to endow nature with godlike status, as an entity that nurtures us if only we will live in harmony with the other species and not demand so much for ourselves.

But nature gives us very few directly usable machine energy resources. Resources are not *taken* from nature, but *created* from nature. What applies to the raw materials of coal, oil, and gas also applies to every raw material in nature—they are all *potential* resources, with unlimited potential to be rendered valuable by the human mind.

Ultimately, a resource is just matter and energy transformed via human ingenuity to meet human needs. Well, the planet we live on is 100 percent matter and energy, 100 percent potential resource for energy and anything else we would want. To say we've only scratched the surface is to significantly understate how little of this planet's potential we've unlocked. We already know that we have enough of a combination of fossil fuels and nuclear power to last thousands and thousands of years, and by then, hopefully, we'll have fusion (a potential, far superior form of nuclear power) or even some hyperefficient form of solar power.

The amount of raw matter and energy on this planet is so incomprehensibly vast that it is nonsensical to speculate about running out of it. *Telling us that there is only so much matter and energy to create resources from is like telling us that there is only so much galaxy to visit for the first time. True, but irrelevant.*

Sustainability is not a clearly defined term. According to the United Nations, it has over a thousand interpretations, but the basic idea is "indefinitely repeatable."[6] For example, the idea of renewability, which is usually synonymous with sustainability in the realm of energy, is that the fuel source keeps replenishing itself over and over without the need to do anything different.

But why is this an ideal? In most realms, we accept and desire *constant change.* For example, you want the best phone with the best materials, regardless of whether those materials will be there in two hundred years and regardless of whether it would be more "renewable" to use two cups and a string.

Why should we want to use solar panels or windmills over and over (leaving aside the fact that they quickly deteriorate and thus require a continuous series of mass-mining projects) if they keep giving us expensive, unreliable energy? Why not use the best, the most *progressive* form of energy at any given time, recognizing that this will change as we advance and the best becomes better?

At the beginning of this book, we observed that human beings survive by using ingenuity to transform nature to meet their needs—i.e., to produce and consume resources. And we observed that the motive power of transformation, the amplifier of human ability, the resource behind every other resource, is energy—which, for the foreseeable future, means largely fossil fuel energy. There is no inherent limit to energy resources—we just need human ingenuity to be free to discover ways to turn unusable energy into usable energy. This opens up a thrilling possibility: the *endless potential for improving life through ever growing energy resources helping create ever growing resources of every kind.* This is the principle that explains the strong correlation between fossil fuel use and life expectancy, fossil fuel use and income, fossil fuel use and pretty much anything good: human ingenuity transforming potential resources into actual resources—including the most fundamental resource, energy.

Growth is not unsustainable. With freedom, including the freedom to produce energy, it is practically inevitable. We are not eating the last slice of pizza in the box or scraping the bottom of the barrel; we are standing on the tip of an endless iceberg.

This is a thrilling prospect for everyone in the world—and certainly for those who live in resource poverty. And if we keep creating resources, I think my future grandchildren will think of my generation of Americans in 2014 as having lived in resource poverty.

HOW USING FOSSIL FUELS ADVANCES FUTURE GENERATIONS

Let's apply the idea of resource creation to the concerns that today's activities are harming future generations, whom the opponents of fossil fuels often focus on.

I do not have children myself, so I know I cannot relate to the perspective on the future that parenthood gives you, but I think it's important for me to try to. So when I think about these issues, I try to think about children I know, children of my close friends or family members, and ask myself how what I'm doing will affect them.

The other day, I was visiting two of my best friends, who have a two-year-old named Seth. I've been fortunate enough to see him about once a week from the time he was born—and it's rare that seeing him is not a highlight of my week. If only we adults could pack as much learning and joy into our lives as a happy two-year-old. Lately, when I visit, I've been teaching him some rudimentary Brazilian jujitsu, because I love it and think it's a great thing for a kid to learn, and I hope that he'll pursue it later in life.

To anyone who has ever connected with even one child, the thought of knowingly taking actions to hurt his future is horrifying. We naturally want our children's lives to be as good as, or better than, ours.

When I see Seth, I sometimes ask myself, *What will he be like in the future?*

Part of the answer to that question is wonderfully unknowable. What choices will he make? I don't know, but I am excited to see.

Part of the answer to that question is within his parents' control: How will his parents influence and educate him, for good or bad? There, I am happy to know that he is in good hands.

But part of the answer to that question is in the control of the rest of us. What choices will *we* make that define the world that he lives in? Will it be a world with more opportunities and fewer hardships or more hardships and fewer opportunities? Will it be a world of progress—a world where he has more exciting career options, less chance of getting sick, more financial security, less chance of going to war, more opportunities to see the world, less suffering, and a cleaner, safer environment? Or will it be a world gone backward, where some or all of these factors get worse?

Everything I've learned about energy has led me to the conclusion that it will be a world of progress if we eagerly pursue more energy, especially fossil fuels, but it will be a world gone backward if we pursue less, out of fear of the environment and climate, which fossil fuels actually make better, not worse.

The basic principle espoused in this book is that we survive by transforming our environment to meet our needs. We maximize resources and we minimize risks. Energy use is the ultimate form of transformation—because it increases our ability to transform our environment to meet every other need, to maximize every resource and minimize every threat.

There is no limit to the amount of resources we can create or the number of problems we can solve—except for the amount of time we have, time being our most valuable resource (though it, too, can be expanded). The only other "limit" of sorts is our starting point—that is, what existing resources we have to work with and, even more important, what *knowledge* we have about resource creation.

What Seth needs is a world where people have created a lot of resources, which will make it easier for him and the others of his generation to create new ones, and a lot of knowledge of how to create resources. I am confident he will get this world, because that's exactly what my generation needed—and got.

Think about your generation. From the perspective of previous generations, you are a future generation. To the extent our grandparents, great-grandparents, and great-great-grandparents thought about what kind of world they would leave, they were thinking about us.

What actions of theirs—and generations before them—benefited us most?

One type of action that benefits everyone going forward is the formation of an important new idea—whether a scientific discovery, such as Newton's three laws of motion, or a technological achievement, such as Watt's efficient steam engine.

If we look at history, an incredibly disproportionate percentage of valuable ideas have come in the last several centuries, coinciding with fossil-fueled civilization. Why? Because such a productive civilization buys us time to think and discover, and then use that knowledge to become more productive, and buy more time to think and discover. We should be grateful to past generations for producing and consuming fossil fuels, rather than restricting them and trying to subsist on something inferior.

If we slow down our progress, including the generation of new ideas, by using inferior energy, we deserve nothing but contempt from future generations—for example, from those who die prematurely because a medical cure comes twenty years later than it needed to.

The production of energy increases the production of knowledge, and it is knowledge that enables one generation to begin where the last left off.

Besides our ideas and knowledge, another form of past action that benefits us is past wealth creation.

Imagine that we had all the knowledge we do today but we were placed in a precivilization environment. By popular accounts, this is a state "rich in natural resources." Would we want to be there? Of course not, because those "resources" would not be genuine resources; they would be only potential resources, raw materials, and it would take a tremendous amount of time and effort to even start using them to create wealth.

The more resources that have been created in the past, the more prosperous societies have been, the more resources they leave behind for us to start with. How grateful am I to the man who first took a streak of rust from a rock and turned it into iron ore, instead of letting it sit there for me and my generation.

That process of resource creation provides the material for the next stage of resource creation. It means taking iron ore and turning it into something more valuable, steel, then taking that steel and turning it into something more valuable, a bunch of girders, then turning those into something much more valuable, a skyscraper, which becomes even more valuable as the workplace for thousands and thousands of productive people, who increase the value of each of those workplaces by starting any number of productive enterprises, which ultimately go back to taking raw materials and making them more valuable through an ingenious combination of machine power, manpower, and superior methods.

Life can be great, indefinitely. Each of us must try to make the best of his life, by creating as much as he wants to benefit his life, and to take joy in the fact that his interests are harmonized with those of his fellow men and his children and his children's children, knowing that the greatest gift he can give to both himself and to the future is to be a creative human being who enjoys his life.

The final point to make about consumption and efficiency and waste is that the most valuable thing we have is our time. If we want to talk about a resource, if human life is our standard, then the most important resource we should be focused on is our time. Using fossil fuels buys us time. It buys us more life. It buys us more opportunities. It buys us more resources. Fossil fuels are an amazing tool with which to create this ultimate form of wealth, this supreme resource: time to use our minds and our bodies to enjoy our lives as much as possible.

Time, and the quality of the life we can enjoy in that time, is already less than it should be, and is threatened to become far, far less than it should be, because even though using fossil fuels is moral, our society does not know it. The voices guiding our society have convinced many of us that the energy of life is immoral and are calling for restrictions that, from all the evidence we have, would be a nightmare.

9

WINNING THE FUTURE

"WHY ARE YOU DOING THIS?"

November 5, 2012

I am standing at the front of a lecture hall packed with three hundred students at the Levine Science Research Center at Duke University. Twelve years ago, I was sitting in one of those seats. Now I'm standing in front, across from Bill McKibben, who has been called the nation's leading environmentalist and is arguably the world's leading opponent of fossil fuels.[1]

Tonight he and I will be debating "the ethics of fossil fuel use." I am nervous. Every debate where there's a camera is immortalized, and no matter how experienced you are, something can always go wrong. Your opponent can make an unexpected point or simply be a bully, and one second of losing your cool can negate an hour and a half of calmness.

McKibben is much more experienced in debates than I am. I've seen him debate before, and he's very skilled—he's calm, thinks well on his feet, and can, on demand, cite any one of hundreds of prestigious studies and authors to make his points.

And he's going to be arguing a much more popular position than I am—that our use of fossil fuels is immoral, an unsustainable "addiction" that causes catastrophic climate change, pollution, and resource depletion. I'm guessing the students are steeped in these ideas; I certainly was when I went to Duke.

I'm going to be arguing that our use of fossil fuels is moral and should be continued and even expanded. I'm going to be arguing that fossil fuels actually *improve* our environment, which is a counterintuitive idea that most people have never heard, one that even I didn't hold a decade earlier.

Preparing for this debate has been rough. Although I am told I am a good debater and many fans of my work have been "talking trash" on my behalf, I have known since day one that I would be in for a war. In the previous two months, I have spent over half my time preparing for this one night.

Usually when I spend the better part of two months on a project, there is at least solace that I am getting paid. Not here. In fact, I am *paying* to do this debate. I have agreed to pay McKibben ten thousand dollars of my own money—which is a lot of money for me, perhaps a reckless amount of money, given that I run a small business, four of five of which fail in their first five years.

What about the deep pockets of my friends in the fossil fuel industry? It is 2012, and I have no such friends. I know barely anyone in the industry, and after months of putting out feelers for some kind of involvement, I managed to get one organization to give me use of their video production crew to film the event. All the promotion and logistics were funded by me, with twenty-five thousand dollars I raised in crowd-funding to promote the debate—overwhelmingly donated by people like me who are outside the fossil fuel industry but value our fossil-fueled civilization—and by our host, the Program for Values and Ethics in the Marketplace at Duke University, run by my former professor Gary Hull.

As I stood onstage, feeling a combination of adrenaline, nerves, and fatigue, I asked myself the question that many of the audience members must have been asking inside their heads: *Why are you doing this?*

It had all started in *Rolling Stone.*

In July 2012, *Rolling Stone* published a fantastically popular piece by Bill McKibben about the evils of fossil fuels entitled "Global Warming's Terrifying New Math." He argued that it had long since been scientifically proven that we needed to restrict the vast majority of our fossil fuel use—at various times, he has called for outlawing between 80 percent and 95 percent over the next several decades. The only thing stopping us, he said, is the political manipulations of the fossil fuel industry, which has fought for its profits at the expense of our future and has thus become "Public Enemy Number One to the survival of our planetary civilization."[2] He called for *a mass-movement to demonize the fossil fuel industry* and deprive it of political standing, much as South Africa's apartheid regime had been demonized and dismantled due to the moral outrage of private citizens around the world. And just as one powerful mechanism for bringing down that regime had been divesting—withdrawing investments from—South African businesses, McKibben called for Americans to divest from the fossil fuel industry as a form of public ostracism.

McKibben's article was a sensation. It received 120,000 "Likes" on Facebook—which an exultant Center for American Progress blogger described as "monster social media numbers of the kind usually reserved for pieces on HuffPost about Kim Kardashian in a bikini."[3] And it was celebrated by citizens and intellectual elites alike.

What did not happen was opposition—least of all by the supposedly big and powerful fossil fuel industry. Was this because they do not fear McKibben? Hardly. McKibben is a master political activist, widely credited with the more than five-year delay of the Keystone XL pipeline, the most prominent fossil fuel project of the last ten years.[4]

The lack of response was, I believed, because McKibben was making a *moral* argument—that it was time to do the *right* thing about fossil fuels for our future, even if it was difficult. And very few people knew that there was a moral argument for fossil fuels, an argument that using them is best for human life across the board, economy and environment, present and future.

Someone had to do something, and I had control of exactly one person. I felt that my best shot at making a difference, even though it was scary and risky, was getting the top anti–fossil fuel advocate on video being challenged on moral grounds. At least then people could see that there was an alternative.

That's why I was there. That's why the stress and the time and the money were worth it. I won't say who I thought won; you can decide for yourself at www.moralcaseforfossilfuels.com. But I was gratified that many people who had never heard the moral case told me they thought it was interesting and important—including a book agent, Wes Neff, who watched it and told me I needed to write the book you're reading right now.

If, at the beginning of this book, it seemed crazy for a human being to invest his life championing fossil fuels and the fossil fuel industry, I hope I have convinced you why this is a more than worthy cause, a cause I hope you'll want to join as part of the broader cause of human flourishing and human progress. Because I am going to ask you, too, to take action, and it won't always be easy and comfortable, but it will be the right thing to do for the present and especially for our future.

As you read this, there are millions of people in the fossil fuel industry working to produce more energy to give us more ability to flourish, but their freedom to produce energy and our freedom to use it are in jeopardy. And as you read this, there is a real, live, committed movement against fossil fuels that truly wants to deprive us of the energy of life. That movement is named the Green movement. To understand how to defend fossil fuels, we must understand the attack, who is attacking, why, and how.

THE ATTACK ON OUR FOSSIL FUTURE

Around the world, in whatever country you live in and maybe the state or city you live in, there are opportunities to produce fossil fuel energy—and almost without exception there are forces conspiring to stop it or that have already stopped it.

In California, where I live, we have perhaps the greatest oil opportunity in the United States, called the Monterey Shale. The companies who want to produce oil there say it has the potential to produce billions of barrels of oil—that's hundreds of billions of gallons of the most coveted energy source on Earth, in a state with arguably the biggest economic problems in the United States.[5] Others, including our Energy Information Administration, claim that the deposit has much less potential, but then, many great oil formations, from Saudi Arabia to North Dakota, were once believed to have no potential. I say, let companies find out. There is no movement in support of the Monterey Shale—but there is a massive movement against it.

This is the story throughout the United States, where shale energy technology, in my view the most exciting technology of our generation, has already been the biggest boon to our economy in the last ten years, and it's just getting started. Or maybe it's just getting stopped. In New York, it's stopped by a moratorium.[6] As I write this, the citizens of Colorado are seriously considering banning it throughout the state.[7] In California, I'll fight for it, but it may well get stopped here.[8]

Or take the coal industry. The United States has been called the Saudi Arabia of coal. Coal is the one source of energy we can be certain can provide energy for as many people as necessary—there are over three thousand years of recoverable reserves at current usage rates.[9] With the right economics and technology, it can be converted into oil fuels, gas fuels, plastics, et cetera. Just consider that. With the right infrastructure, this is a source of energy that we *know* could take billions of people from not being able to power a fan to cool them down or a radiator to keep them warm all the way to central air-conditioning and heating. Not that it would be easy—energy is just one part of development, which requires rule of law and economic freedom, among other things—but we can count on the coal industry to produce all the energy that is needed.

Shouldn't that inspire support?

Culturally, it never seems to. Coal is "dirty," as if it's the only energy source that has risks and side effects, and as if it doesn't have the benefit of being the hope for billions of people to have a better, healthier life.

In the United States, we have a place in Wyoming called the Powder River Basin, one of the greatest coal deposits of all time, mined with state-of-the-art mining technology that can extract more coal from a mine than ever before.[10] But the participants in the project, such as Peabody Energy, have had to fight daily for permission to empower billions of people. And in particular, they have to fight for permission to export that coal in the Pacific Northwest.[11] At the start of 2013, there were six export project proposals in the Pacific Northwest. By the end of the year, three of the six had been dropped due to fierce opposition from environmentalist organizations.

The message of those organizations is summed up by Robert F. Kennedy Jr.: "They're coming to ship their poison so they can poison the people in China. And that poison's going to come back here and poison your salmon and your children, so don't let it happen."[12] This "poison" is the basis of life-giving energy technology that has given Chinese people years more of health.

In place after place, the energy of life is portrayed as deadly, its producers immoral. Why?

If you agree with me that using fossil fuels is a moral imperative, that more energy is more ability and the only way 7 billion people are going to get it anytime soon is with more fossil fuels, then I hope you want to fight for fossil fuel freedom around the world. But to do that, we need to understand why that cause is losing, why our culture as a whole believes, despite the evidence, that fossil fuels are not a healthy, moral choice but a dangerous, immoral addiction.

In one sense, the answer to "Why do we believe the wrong thing about fossil fuels?" is simple. Lack of education. We haven't been taught all the right facts. We aren't taught in school how energy makes our climate safer, only how CO_2 emissions supposedly make it more dangerous. We aren't taught in school how energy makes our environment better, only ways (usually exaggerated) in which fossil fuels make it dirtier. We aren't taught in school how the fossil fuel industry is a resource-creating industry; we are taught that it is shamelessly exploiting dwindling natural resources. If only the truth were taught, the world would be a different place, right?

In a sense, yes—but that raises a deeper question: Why are we as a culture so oblivious to the facts? In particular, why are we so oblivious to *positive* facts about fossil fuels and so susceptible to negative *fabrications* about fossil fuels? We are surrounded by a better and safer world that runs on fossil fuel energy. Why can't we see it?

Here's my answer: The reason we have come to oppose fossil fuels and not see their virtues is not primarily because of a lack of factual knowledge, but because of the *presence of irrational moral prejudice* in our leaders and, to a degree, in our entire culture.

Anytime someone is oblivious to the positive and inclined toward the negative, he has a *prejudice*. Consider racial prejudice. Someone with, say, a racial prejudice against blacks will tend to ignore the virtues of a black individual he meets and exaggerate (or manufacture) vices.

There is clearly a prejudice in how our culture processes information about fossil fuels. Unless we understand and correct the source of that prejudice, factual education will be an uphill battle.

The prejudice, which is held consistently by our environmental thought leaders and inconsistently by the culture at large is the idea that *nonimpact on nature is the standard of value.* It is better known by a single color: Green.

UNDERSTANDING THE ANTI-FOSSIL FUEL MOVEMENT

In this book, I have quoted extensively from certain environmental thought leaders—Paul Ehrlich, Al Gore, Bill McKibben, Amory Lovins, John Holdren—precisely because they are thought *leaders:* They have had tremendous influence throughout the culture.

We've seen that these thought leaders not only come to certain deadly *conclusions* and *policies,* but also keep using the same faulty *method of thinking:* they exaggerate the negatives of fossil fuels and ignore or greatly understate the positives.

Notice that they, and practically all other environmental or Green leaders, express little enthusiasm for the value of cheap, plentiful, reliable energy or the unique ability of the fossil fuel industry to provide it; instead, they keep claiming, without evidence, that expensive, unreliable, unscalable energy will somehow become cheap, reliable, and scalable—unconcerned by what will happen if and when they are wrong and nothing can make up for the energy they've taken away from us. They also cannot see much positive in nuclear or hydro. They claim to care about climate but are indifferent to the climate mastery fossil fuels give us and the opportunity to give that mastery to billions more, but they will use every fallacy in the book to make us terrified of fossil fuel–related climate change. They claim to care about a clean environment but have nothing but scorn for the industry that gave us the ability to create the cleanest, healthiest environment in history. They claim to care about abundant resources, but are indifferent to the fact that the fossil fuel industry is itself producing new resources and helping every other industry produce new resources—and that restricting fossil fuel use would bring us that much closer to the resource poverty that has been mankind's condition for all but a recent sliver of history.

Thought leaders are usually extremely bright men and women, and all of these thought leaders are bright. At the same time, all of them have been confronted, in one way or another, with the data I have presented in this book. Yet they still say fossil fuels are catastrophic and seem to have absolutely zero fear of the nearly infinite risk of *not* using fossil fuels at this stage of history.

Why?

It goes back to the issue of *standard of value.*

The environmental thought leaders' opposition to fossil fuels is not a mistaken attempt at pursuing human life as their standard of value. They are too smart and knowledgeable to make such a mistake. Their opposition is a *consistent* attempt at pursuing their actual standard of value: a pristine environment, unaltered nature. Energy is our most powerful means of transforming our environment to meet our needs. If an unaltered, untransformed environment is our standard of value, then *nothing could be worse than cheap, plentiful, reliable energy.*

I'm saying that if fossil fuels created no waste, including no CO_2, if they were even cheaper, if they would last practically forever, if there were no resource-depletion concerns, the Green movement would still oppose them.

This is hard to believe. Which is why you need to know the following story.

For many decades, the ultimate energy fantasy has been what's called nuclear fusion. Conventional nuclear power is called nuclear *fission,* which unleashes power through the decay of heavy atoms such as uranium. Nuclear fusion unleashes far more power through fusion of two light atoms, of hydrogen, for example. Fusion is what the sun uses for energy. But all human attempts at fusion so far have been inefficient—they take in more energy than they produce. But if it could be made to work, it would be the cheapest, cleanest, most plentiful energy source ever created. It would be like the problem-free fossil fuels I said the Green leaders would oppose.

In the late 1980s, some reports that fusion was close to commercial reality got quite a bit of press. Reporters interviewed some of the world's environmental thought leaders to ask them what they thought of fusion—testing how they felt, not about energy's human-harming risks and wastes but its pure transformative power. What did they say?

There are some quotes from a story in the *Los Angeles Times* called "Fear of Fusion: What if It Works?"

Leading environmentalist Jeremy Rifkin: "It's the worst thing that could happen to our planet."[13]

Paul Ehrlich: Developing fusion for human beings would be "like giving a machine gun to an idiot child."[14]

Amory Lovins was already on record as saying, "Complex technology of any sort is an assault on human dignity. It would be little short of disastrous for us to discover a source of clean, cheap, abundant energy, because of what we might do with it."[15]

He is talking here about something that, if it had worked, would have been able to empower every single individual on the globe and that undoubtedly would have given *him* a longer life through the increased scientific and technological progress a fusion-powered society would make. He's talking about something that could take someone who had never had access to a lightbulb for more than an hour, and give him all the light he needed for the rest of his life. He's talking about something that would have given that hospital in

The Gambia the power that it needed to save the two dead babies in the story, who could have been thriving eight-year-olds as I write this, instead of painful memories for would-be parents.

That is what Amory Lovins regards as disastrous "because of what we might do with it."[16] Well, we've seen what we do with energy—we make our lives amazing. We go from physically helpless to physical supermen. We build skyscrapers and hospitals. We take vacations and go on honeymoons. We visit our families and tour the world. We relieve drought and vanquish disease. We transform the planet for the better.

Better—by a human standard of value.

But if your standard of value is unaltered nature, then Lovins is right to worry. With more energy, we have the ability to alter nature more, and we will do so—because transforming our environment, transforming nature, is our means of survival and flourishing.

To the antihumanist, that's precisely the problem. Have you ever heard mankind described as a cancer on the planet? Prince Philip, former head of the World Wildlife Fund, has said, "In the event that I am reincarnated, I would like to return as a deadly virus, in order to contribute something to solve overpopulation."[17] Remember that in chapter 1, David M. Graber, in praising the theme of Bill McKibben's book, said, "Until such time as *Homo sapiens* should decide to rejoin nature, some of us can only hope for the right virus to come along."[18]

This is the logical end of holding human nonimpact as your standard of value; the best way to achieve it is to do nothing at all, to not exist. Of course, few hold that standard of value consistently, and even these men do not depopulate the world of themselves. But to the extent that we hold human nonimpact as our standard of value, we are going against what our *survival requires.*

And our culture has accepted this toxic standard in large doses under the friendly label "Green."

OUR PREJUDICED CULTURE

In the last section, about the thought leaders, I observed that on every single issue pertaining to fossil fuels they would greatly exaggerate the negatives of fossil fuels and ignore or greatly understate the positives.

But let's focus now on our culture. How different are we from the thought leaders who influence our culture? I think our motives are much better, but we have adopted many of their same bad thinking methods, and we partially share their nonimpact standard of value.

Notice that, with each issue surrounding fossil fuels, we all too easily believe the negatives and are blinded to the positives. How many of us have ever thought to appreciate the man-made miracle that is cheap, plentiful, reliable energy?

How many of us appreciate the people who actually produce it, rather than demonize them and laud their imaginary replacements in the solar and wind industries?

How many of us consider the *possibility* that human beings could be a *positive* force climatewise, whether by fertilizing the atmosphere or by creating an environment that maximizes climate benefits and minimizes climate risks?

How many of us consider the possibility that we are improving our environment by using fossil fuels? In my experience, not even the fossil fuel industry considers that possibility.

As a culture, we are consistently inclined to view the fossil fuel industry as negative, and in particular, *environmentally* negative.

Why? Because we haven't been taught the facts? That doesn't explain it—why don't we look for positive environmental facts about the fossil fuel industry, instead of *assuming that they don't exist*? Because we believe that to be environmentally good, to follow an environmentally good standard of value, is to be "green," to not have an *impact* on things.

Green is often associated with a lack of pollution and other environmental health hazards, but this is both far too narrow and highly misleading. Consider the range of actions that fall under the banner of Green. It is considered Green to object to crucial industrial projects, from power plants to dams to apartment complexes, on the grounds that some plant or animal will be affected, plants and animals that take precedence over the human animals who need or want the projects.

It is considered Green to do less of anything industrial, from driving to flying to using a washing machine to using disposable diapers to consuming pretty much any modern product (there is now an attack on iPhones for being insufficiently Green, given the various materials that must be mined to make them).

The essence of "going Green," the common denominator in all its various iterations, is the belief that *humans should minimize their impact on nonhuman nature.*

The difference between our culture and the Green movement is that our culture believes that you can't always be environmentally good; our culture regards Green as one of many competing ideals that we must balance. But this attempt to balance being on a human standard of value sometimes and a nonhuman standard at other times is like trying to create a balanced diet that includes food and poison.

Why do we accept the Green ideal, the ideal that causes us to hate our greatest energy technology and the people who produce it? In large part, we do so because environmental leaders have made us associate the antihuman ideal of nonimpact with something very good: minimizing pollution, that is, minimizing negative environmental impacts. But if you're antipollution, Greenness or nonimpact is a confusing and dangerous way of thinking about the issue, for by associating impact with something negative, you're conceding that *all* human impact is somehow bad for the environment.

And that's what the Green movement wants you to believe.

Instead of recognizing that transforming our environment is a life-serving virtue that can have environmentally undesirable risks and side effects, the Green movement wants you to look at *all* transformation of our environment as environmentally bad.

In fact, the *worst* thing we can do environmentally is *not* transform our environment, because then we would live with the threat-laden and resource-poor environment of undeveloped nature.

Another reason we buy into Green is because we as a culture have never been fully comfortable with human industry. We're taught that the pursuit of profit is wrong, that capitalism is wrong, and that we should feel guilty for our wealth and way of life.

Accepting nonimpact as our environmental ideal primes us to swallow any argument that an industry's environmental impact is too high and to assume that the consequences of any environmental impact *must be bad*—even while we wake up every day in the greatest environment in history.

That's the power of prejudice—prejudice that comes from holding a false philosophy we don't know we accept and that most of us would fully reject if we saw its real meaning.

Now that we know its meaning, we can look for—and embrace—a new, prohuman environmental philosophy.

A NEW IDEAL: INDUSTRIAL PROGRESS

So long as we accept nonimpact as an environmental ideal, we will not fight passionately against those who oppose the energy of life, because we won't consider its essence—the transformation of nature in service of human life—as a moral ideal.

But transformation *is* a moral ideal. I call that ideal *industrial progress*—the progressive improvement of our environment using human industry, including energy and technology, in service of human life. It's why I named my think tank the Center for Industrial Progress. I wanted to start a positive alternative to the mainstream Green environmentalist movement, to replace the deadly ideal of nonimpact with the true ideal of industrial progress. We don't want to "save the planet" *from* human beings; we want to *improve* the planet *for* human beings.

We need to say this loudly and proudly. We need to say that human life is our one and only standard of value. And we need to say that the transformation of our environment, the essence of our survival, is a supreme virtue. We need to recognize that to the extent we deny either, we are willing to harm real, flesh-and-blood human beings for some antihuman dogma.

Making a moral case always means naming your standard—for us, human life. It means tying everything, including every positive and negative of fossil fuel use, to human life. If you do that in your thinking, I believe you will come to conclusions similar to mine. If you do that in communicating with others, you will be amazingly effective, because you will be clear and sincere.

If we can do this, we can create the dream—an energy revolution that spawns revolutions in every other field. And we can perform a great act of justice for the millions of men and women in the fossil fuel industry who have been working every day to keep our machines alive, who have been given little appreciation by our culture but much condemnation, and who in my experience do not themselves understand the full importance of their work. I hope this book helps them see it.

The fossil fuel industry is a *moral* industry at its core. Members do immoral things, to be sure, but transforming ancient dead plants into the energy of life in a way that maximizes benefits and minimizes risks is an activity that the industry should be proud of, and we should be proud to use its product.

Unfortunately, the fossil fuel industry has, in recent decades, not believed that or at least has refused to say it. It has conceded to its environmentalist opponents that fossil fuels are an "addiction," just a *temporarily necessary* one.

Last year, I wrote an open letter to executives in the fossil fuel industry criticizing them for this conduct and asking them to join me in making a moral case for their industry. I want to include a shortened version of it in this book, because what they say about their industry doesn't affect just them; it affects all of us.

WHAT THE FOSSIL FUEL INDUSTRY MUST DO

To leaders of the fossil fuel industry:

Here's a typical communications plan of yours to win over the public.

- "We will explain to the public that we contribute to economic growth."
- "We will explain to the public that we create a lot of jobs."
- "We will link our industry to our national identity."
- "We will stress to the public that we are addressing our attackers' concerns—by lowering the emissions of our product."
- "We will spend millions on a state-of-the-art media campaign."

Why doesn't it work? Well, imagine if you saw the same plan from a tobacco company. It would tie increased tobacco sales to economic growth, to job creation, to national identity, to reducing tar. Would you be convinced that it would be a good thing if Americans bought way more tobacco?

I doubt it, because none of these strategies does anything to address the industry's fundamental problem, the fact that use of the industry's core product, tobacco, is viewed as a self-destructive addiction. So long as that is true, the industry will be viewed as an inherently immoral industry. And so long as *that* is true, no matter what the industry does, its critics will always have the moral high ground.

You might say that it's offensive to compare the fossil fuel industry to the tobacco industry—and you'd be right. But in the battle for hearts and minds, you are widely viewed as worse than the tobacco industry.

Your attackers have successfully portrayed your core product, fossil fuel energy, as a self-destructive addiction that is destroying our planet, and characterized your industry as fundamentally immoral. In a better world, the kind of world we should aspire to, they argue, the fossil fuel industry would not exist.

There is only one way to defeat the environmentalists' moral case against fossil fuels—refute its false central idea that fossil fuels destroy the planet. If we don't refute that idea, we accept it, and if we *accept that fossil fuels are destroying the planet,* the only logical conclusion is to cease new development and slow down existing development as much as possible.

Unfortunately, the fossil fuel industry has not refuted the moral case against fossil fuels. In fact, the vast majority of its communications *reinforce the moral case against fossil fuels.*

For example, take the common practice of publicly endorsing renewables as the ideal. Fossil fuel companies, particularly oil and gas companies, proudly feature windmills on Web pages and annual reports, even though these are trivial to their bottom line and wildly uneconomic. This obviously implies that renewables are the goal, oil and gas being a temporarily necessary evil.

Another way in which the fossil fuel industry reinforces the moral case against itself is by trying to sidestep the issue with talk of jobs or economics or patriotism. While these are important issues, it makes no sense to pursue them via fossil fuels if they are destroying our planet. That's why environmentalists compellingly respond with arguments such as: Do we want economic growth tied to poison? Do we want more jobs where the workers are causing harm? Do we want our national identity to continue to be associated with something we now know is destructive?

There are many, many more forms of conceding the environmentalists' moral case and giving them the high ground. Here are half a dozen more just to give you a sense of the scope of the problem.

- *Not mentioning the word "oil"* on home pages (this has at times been true of ExxonMobil, Shell, and Chevron). This implies that you're ashamed of what you do and that your critics are right that oil is a self-destructive addiction.
- Focusing attention on everything but your core product—community service initiatives, charitable contributions, et cetera. This implies that you're ashamed of your core product.
- Praising your attackers as "idealistic." This implies that those who want your destruction are pursuing a legitimate ideal.
- Apologizing for your "environmental footprint." This implies that there's something wrong with the industrial development that is inherent in energy production.
- Spending most of your time on the defensive. This implies that you don't have something positive to champion.
- *Criticizing your* opponents primarily for getting their facts wrong without refuting their basic moral argument. This implies that the argument is right but your opponents just need to identify your evils more precisely.

The industry's position amounts to this: "Our product isn't moral, but it's something that we will need for some time as we transition to the ideal fossil fuel–free future." What you're telling the world is that you are a necessary evil. And because the environmentalists agree that it will take some time to transition to a fossil fuel–free future, the argument amounts to a debate over an expiration date. Environmentalists will argue that fossil fuels are necessary for a shorter time, and you'll argue that they're necessary for a longer time, so they'll always sound optimistic and idealistic, and you'll always sound cynical and pessimistic and self-serving. So long as you concede that your product is a self-destructive addiction, you will not win hearts and minds—and you will not deserve to.

In my experience, whatever the audience and whatever the medium, making the moral case for the fossil fuel industry is a game changer. We need you to make that case—for your sake and ours. I believe that if enough of us work together applying these ideas, the unimaginable is possible. In the future, I see:

- Pro–fossil fuel politicians winning spectacular victories over anti–fossil fuel politicians in debates.
- Energy companies having inspiring, iconic campaigns that make them as cool as iPhones.
- Workforces full of incredibly educated, motivated, articulate ambassadors.
- Associations training members in moral communication.
- News stories with quotes by morally confident, persuasive CEOs.
- Web sites having more emotional resonance than the Greenpeace or Sierra Club Web sites.
- Widespread criticism of anyone who delays a pipeline for five years, not as proenvironment but as antiprogress.
- A new generation of intellectuals who are passionate advocates of fossil fuels.

- College campuses where students are not afraid to say, "I love fossil fuels."

I wrote this letter as part of an effort of mine in the last several years to convince the fossil fuel industry to make a moral case for its work. Historically, it has been a major bankroller of Green organizations. For example, between 2007 and 2010, the natural gas industry gave $25 million to the Sierra Club.[19] I've told them, in effect, "How am I supposed to fight for the freedom to use your product if you won't?" I'm happy it's started working, and I now spend quite a bit of my time working with companies to improve their communications. It's in their self-interest, because they can get projects approved quickly if they actually explain the value of what they do. For me, the prospect of getting the resources of the industry deployed to actually make a compelling case is beyond exciting. Because I am a capitalist and charge for my services, maybe I will get attacked now for being paid by the fossil fuel industry. But there's the prejudice again. Why would someone assume that someone who works with fossil fuel companies is corrupt, while those who, say, accept government grants aren't? As I advise members of the industry on what to say, I don't say this industry is good because I work with it; I work with this industry because I think it's good.

The fossil fuel industry has a giant megaphone it can use to influence for good or bad. It's in everyone's interest for it to use it for good.

WHAT WE ALL MUST DO

In 2007 and 2008, candidate Obama declared his intention to destroy fossil fuel energy in America and around the world, calling for "emissions targets" that would make it illegal to use more than 20 percent of today's levels.[20] About oil, the most versatile fuel in the world, which powers 93 percent of our transportation system and, through shale-oil booms in North Dakota, Texas, and elsewhere, has been one of our few sources of economic hope, he said:

> At the dawn of the twenty-first century, the country that faced down the tyranny of fascism and communism is now called to challenge the tyranny of oil. . . . For the sake of our security, our economy, our jobs and our planet, the age of oil must end in our time.[21]

While he was saying this, the oil industry he was comparing to the mass murderers of the twentieth century was perfecting shale-oil (and shale-gas) technology. Thanks to Obama's lack of oversight in this area, shale energy technology became the leading positive economic force during his administration.

That is, *a revolution in fossil fuel technology occurred because our government didn't know enough about it to demonize and ban it.* This is not the kind of thing we want to depend on.

What if Obama had been aware of this revolution in the making ten years out? He would have no doubt regarded it as a dangerous practice to be stopped, given that he viewed oil as a "tyranny" to be ended, not expanded. Technological progress in the United States would have been thwarted—and with it, progress around the world. The United States is the best place in the world to do fossil fuel research and development, because we have the most private property that can be bought and explored, rather than delegated at the whim of the state.

This example to me captures where we are—incredible threats to progress and incredible opportunities for progress. We are still arguably at the beginning of the fossil fuel age. In several decades we may be able to drill efficiently and safely at any depth, efficiently turn coal or gas into oil, and use fossil fuels to help develop new generations of technology (likely nuclear) and help increase the amount of the most valuable resources—food, water, beauty, and most important, human time. All indications are that, as the amount of CO_2 in the atmosphere increases from .04 percent to .05 or .06 percent, we will continue to benefit from more plant growth. If new climate dynamics are discovered, we will adapt—always keeping in mind as full context the indispensable value of industrial civilization.

We *can* have it all.

We just need to be clear on what is right, then take the time and sometimes social risk to try to reach the people who matter most to us. I wrote this book so you could hand it to the people who matter most to you—and so that you could take its ideas and make them your own, telling the people who matter to you how you think and feel.

I wrote this book for anyone who wants to make the world a better place—for human beings—including many, many people who would start this book opposed to or at least suspicious of fossil fuels. Having held that position myself before, I know it can be well motivated. The idea of ruining the world for the less fortunate and, even worse, for our children or grandchildren is horrifying to us. Thus, when someone tells us of a major risk that our behavior is causing, we want to do something about it.

What we are not taught is that the biggest risk is *not* using fossil fuels, and that using them is incredibly virtuous. We are not taught that we're building a civilization that serves us and the future, that we're creating knowledge and resources that can enrich everyone around the world. We're not taught that the choices we make often reflect an extremely rational calculation that balances benefit and risk. We're not taught that some people truly believe that human life doesn't matter, and that their goal is not to help us triumph over nature's obstacles but to remove us as an obstacle to the rest of nature.

Make no mistake—there are people trying to use you to promote actions that would harm everything you care about. Not because they care about you—they prioritize nature over you—but because they see you as a tool.

The unpopular but moral cause of our time is fossil fuels. Fossil fuels are easy to misunderstand and demonize, but they are absolutely good to use. And they absolutely need to be championed.

There are many specific battles to be fought. The venue and strategy for each is ever changing, which is why the specific actions we take need to be timely and coordinated. That's why this book has a Web site, www.moralcaseforfossilfuels.com, which will let you know about the latest opportunities to fight for energy liberation, whether it's promoting a series of debates over fossil fuels, writing a public comment on the EPA's attacks on coal, or sharing inspiring stories about industrial progress around the world.

But no matter what you read, the need for moral clarity will always be timely. Here, in a sentence, is the moral case for fossil fuels, the single thought that can empower us to empower the world: Mankind's use of fossil fuels is supremely virtuous—because human life is the standard of value, and because using fossil fuels transforms our environment to make it wonderful for human life.

ACKNOWLEDGMENTS

One theme of this book is the importance of using experts the right way. If we can use the best experts as advisers and get them to clearly share what they know and how they know it, our understanding will be far better.

That was certainly true for the creation of this book. It was enormously improved by numerous experts and other talented people who helped me in every aspect.

Philosopher Greg Salmieri helped me clarify my thinking about practically every issue in this book, thanks to his ruthlessly precise thinking and original insights about everything from the proper way to deal with experts to how to structure the book.

Steffen Henne, our lead researcher at the Center for Industrial Progress, helped me find the high-quality and wide-ranging data I was looking for. Steffen's breadth of knowledge about practically every topic in this book is incredible, as is his ability to catch errors.

David Epstein, the best expert at synthesizing and graphing data I have ever met, also, fortunately for me, happens to be my father. Almost every image in this book is produced by his proprietary software, G3, which is a stupendously powerful tool for synthesizing a multiplicity of data sets and then displaying them visually. Thanks also to graphic designer Marianne Epstein (another fortuitous relation), who helped with design across the board.

Alex Hrin, my friend and a trained biophysicist who has a wide array of scientific and technological knowledge, helped me greatly in formulating many of the sections on climate and environmental impact.

Don Watkins, my longtime friend and colleague, provided essential editing and feedback at every stage. Don has a remarkable ability to candidly communicate all the problems with my writing but simultaneously motivate me with his enthusiasm about what the final product will be.

Maria Gagliano, my editor at Portfolio, defied all expectations, improving every section and paragraph she commented on by being, as she put it, "your toughest critic." In an age when editors are known to be "light touch," I was lucky to have one who put in the time to make every page better.

I got many helpful comments and ideas from others. CIP Senior Fellow Eric Dennis, a formally trained physicist and self-trained economist, was ever insightful about both climate and economic issues. My friend Chad Morris gave me dozens of suggestions and criticisms, and the book benefited from every one of them. My friend Jesse McCarthy, a master storyteller, helped me make the ideas in this book more powerful and more visual. The members of CIP's "Talent Factory" training program, which includes experts in everything from chemical engineering to petroleum engineering, provided dozens of good ideas and clarifications.

My mentor Dennis Farrier helped me create an overall vision for the book. He, along with Lisa VanDamme, Aaron Briley, Johnny Saba, Ruth Epstein, Robert Bradley Jr., and many other friends and family, gave me continual moral support—as did the thousands of supporters of the Center for Industrial Progress, who are a nonstop source of fuel.

Every project needs a project manager and I am grateful to CIP's Erin Connors for making sure that all the pieces got done when they needed to. Erin's genuine enthusiasm for the project and active desire to make it as good as possible led her to help out in practically every aspect—from conducting research to finding powerful stories to catching unclarity to rigorously making sure every citation was right.

I also want to acknowledge the intellectual influences whose ideas led me to be able to write this book in the first place. This book would not have been possible without clear thinking about standards of value, which I got (among many other philosophical identifications) from Ayn Rand; without a proper understanding of resources, which I learned from the works of economists Julian Simon and George Reisman; and without extensive knowledge of energy, which I obtained first and foremost from the works of Petr Beckmann. Perhaps the person who has helped me the most over the years in thinking through these issues is my former colleague, philosopher Onkar Ghate, who is a master at questioning conventional ways of framing issues and then coming up with far better ones.

Finally, I must thank the person without whom this book would not exist: my agent, Wes Neff. Wes contacted me out of the blue after reading an essay of mine called "The Moral Case for Fossil Fuels," and told me he thought it contained new and important ideas that deserved a hearing—and that he, as president of one of America's leading agencies, would put his time and his name behind. I have never met an agent with his genuine enthusiasm for new ideas. And thanks to his efforts, I got to do this book with Portfolio / Penguin, my number one choice for a publisher. And it was the right choice; from beginning to end and top to bottom everyone at Portfolio has been a pleasure to work with.

Finally, I'd like to thank the people who work in the fossil fuel industry. I came to this industry as a complete outsider—I had been writing about it for years before I knew more than five people in it—but it has been gratifying to meet and even befriend many members of the industry. Every day, you are working to make sure that everything in our lives works. I hope that this book makes more people appreciate that, and I hope that this book makes *you* better able to appreciate that.

SELECT BIBLIOGRAPHY

Boden, T. A., G. Marland, and R. J. Andres, "Global, Regional, and National Fossil-Fuel CO_2 Emissions." Carbon Dioxide Information Analysis Center (CDIAC), Oak Ridge National Laboratory, U.S. Department of Energy, Oak Ridge, Tennessee 2013. doi:10.3334/CDIAC/00001_V2013.

Bolt, J., and J. L. van Zanden, "The First Update of the Maddison Project; Re-Estimating Growth Before 1820." Maddison Project Working Paper 4. Accessed June 8, 2014. http://www.ggdc.net/maddison/maddison-project/home.htm.

BP. Statistical Review of World Energy 2013. "Historical Data Workbook." Accessed May 20, 2014. http://www.bp.com/en/global/corporate/about-bp/energy-economics/statistical-review-of-world-energy.html.

Carbon Dioxide Information Analysis Center. U.S. Department of Energy, Oak Ridge National Laboratory. Oak Ridge, Tennessee. http://cdiac.ornl.gov/ftp/trends/co2/lawdome.smoothed.yr20.

Christy, John C. Climate Model Output from KNMI Climate Explorer using sources referenced in chart. University of Alabama—Huntsville.

"Coal-Fired Power Plant Hamm." ALPINE Bau GmbH. Accessed July 10, 2014. http://www.alpine.at/en/bereiche/kraftwerksbau/steinkohle-kraftwerk-hammsteinkohle-kraftwerk-hamm/.

"Commodity Food Price Index." Index Mundi. Accessed on July 12, 2014. http://www.indexmundi.com/commodities/?commodity=food-price-index.

EM-DAT: The OFDA/CRED International Disaster Database. Université catholique de Louvain, Brussels, Belgium. Accessed May 3, 2014. http://www.emdat.be.

Etheridge, D.M., et al., "Historical CO_2 Records from the Law Dome DE08, DE08-2, and DSS Ice Cores." Trends: A Compendium of Data on Global Change (June 1998).

European Energy Exchange AG, Leipzig, Germany. "Transparency Platform Data." Accessed on July 9, 2014.

Federal Statistical Office of Germany. GENESIS-Online database. "Monatsbericht über die Elektrizitätsversorgung der Netzbetreiber." Accessed July 10, 2014. https://www-genesis.destatis.de.

"Global Tropical Cyclone Accumulated Cyclone Energy." Global Tropical Cyclone Activity. Accessed July 10, 2014. http://models.weatherbell.com/tropical.php.

"Global Tuberculosis Report 2013." World Health Organization. Geneva, Switzerland: WHO Press, 2013. http://apps.who.int/iris/bitstream/10665/91355/1/9789241564656_eng.pdf.

"HadCRUT4." Met Office Hadley Centre observations datasets. Accessed July 5, 2014. http://www.metoffice.gov.uk/hadobs/hadcrut4/.

Hansen, J., et al., "Global Climate Changes as Forecast by Goddard Institute for Space Studies Three-Dimensional Model." *Journal of Geophysical Research* 93 (August 20, 1988): 9341–46. Goddard Institute for Space Studies, National Aeronautics and Space Administration, New York. doi: 10.1029/JD093iD08p09341.

Holgate, Simon J., et al., "New Data Systems and Products at the Permanent Service for Mean Sea Level (PSMSL)." *Journal of Coastal Research* 29, no. 3 (2013), 493–504. doi:10.2112/JCOASTRESD-12-00175.1.

Idso, Craig. "Plant Growth Database." Center for the Study of Carbon Dioxide and Global Change. Accessed May 4, 2014. http://www.co2science.org/data/plant_growth/plantgrowth.php.

Keeling, C. D. et al., "Exchanges of Atmospheric CO_2 and $13CO_2$ with the Terrestrial Biosphere and Oceans from 1978 to 2000." I. Global aspects. SIO Reference Series, No. 01-06. Scripps Institution of Oceanography, San Diego. (2001).

MacFarling Meure, C. et al., "Law Dome CO_2, CH_4 and N_2O Ice Core Records Extended to 2000 Years BP." *Geophysical Research Letters*, Vol. 33. (2006). L14810. doi:10.1029/2006GL026152.

Maue, Ryan N., "Recent Historically Low Global Tropical Cyclone Activity." *Geophysical Research Letters* 38 (July 20, 2011). L14803. American Geophysical Union. doi:10.1029/2011GL047711.

Mears, Carl A., Matthias C. Schabel, and Frank J. Wentz, "A Reanalysis of the MSU Channel 2 Tropospheric Temperature Record." *J. Cli-*

mate, 16 (2003), 3650–3664. doi: http://dx.doi.org/10.1175/1520-0442(2003)016<3650:AROTMC>2.0.CO;2.

"Merged Ice-Core Record Data." Scripps Institution of Oceanography CO_2 Program. Accessed July 9, 2014. http://scrippsco2.ucsd.edu/data/atmospheric_co2.html.

Morice, C.P., et al., "Quantifying Uncertainties in Global and Regional Temperature Change Using an Ensemble of Observational Estimates: The Hadcrut4 Dataset." *Journal of Geophysical Research* (2012), 117, D08101. doi:10.1029/2011JD017187.

Myhre, Gunnar, Eleanor J. Highwood, Keith P. Shine, and Frode Stordal, "New Estimates of Radiative Forcing Due to Well-Mixed Greenhouse Gases," *Geophysical Research Letters* 25, No. 14, pages 2715–18, Jul. 15, 1998, http://onlinelibrary.wiley.com/doi/10.1029/98GL01908/abstract

"National Emissions Inventory (NEI) Air Pollutant Emissions Trends Data." U.S. Environmental Protection Agency. Accessed July 10, 2014. http://www.epa.gov/ttnchie1/trends/.

Peterson, Per F., Haihua Zhao, and Robert Petroski, "Metal and Concrete Inputs for Several Nuclear Power Plants." Report UCBTH-05-001, UC Berkeley. February 4, 2005.

"Progress on Drinking Water and Sanitation." WHO/UNICEF Joint Monitoring Program update 2014. (May 2014.) Accessed July 10, 2014. www.who.int/water_sanitation_health/publications/2014/jmp-report/.

"RSS Lower Troposphere Data." Microwave Sounding Unit Temperature Anomalies. U.S. National Oceanic and Atmospheric Administration, National Climatic Data Center. Accessed on July 9, 2014. http://www.ncdc.noaa.gov/temp-and-precip/msu/.

"Tide Gauge Data." Permanent Service for Mean Sea Level (PSMSL). 2014. Accessed July 10, 2014. http://www.psmsl.org/data/obtaining/.

Wilburn, D.R. "Wind Energy in the United States and Materials Required for the Land-Based Wind Turbine Industry from 2010 Through 2030." U.S. Geological Survey Scientific Investigations Report 2011–5036.

"World Development Indicators." World Bank. Accessed June 8, 2014. http://data.worldbank.org/products/wdi.

NOTES

CHAPTER 1:

THE SECRET HISTORY OF FOSSIL FUELS

1. BP, *Statistical Review of World Energy 2013*, "Historical Data Workbook" [XLSX], June 2013, www.bp.com/en/global/corporate/about-bp/energy-economics/statistical-review-of-world-energy-2013.html.
2. Ibid.
3. Peter Voser, "Getting the Future Energy Mix Right: How the American Shale Revolution Is Changing the World," Shell, speech, Boston, Mar. 21, 2013, www.shell.com/global/aboutshell/media/speeches-and-webcasts/2013/getting-the-future-energy-mix-right.html.
4. Elizabeth Bumiller and Adam Nagourney, "Bush: 'America Is Addicted to Oil,'" *New York Times*, Feb. 1, 2006, www.nytimes.com/2006/02/01/world/americas/01iht-state.html?pagewanted=all&_r=0.
5. Intergovernmental Panel on Climate Change, "Potential of Renewable Energy Outlined in Report by the Intergovernmental Panel on Climate Change," press release, Abu Dhabi, May 9, 2011, http://srren.ipcc-wg3.de/press/content/potential-of-renewable-energy-outlined-report-by-the-intergovernmental-panel-on-climate-change.
6. Justin Gillis and Kenneth Chang, "Scientists Warn of Rising Oceans," *New York Times*, May 12, 2014, www.nytimes.com/2014/05/13/science/earth/collapse-of-parts-of-west-antarctica-ice-sheet-has-begun-scientists-say.html?_r=0; Ben Wolford, "Is It Too Late to Save Our Cities from Sea Level Rises?" *Newsweek*, June 5, 2014, www.newsweek.com/2014/06/13/it-too-late-save-our-cities-sea-level-rise-253447.html.
7. Bill McKibben, "The Ethics of Fossil Fuel Use" (debate at Duke University, Durham, NC, November 5, 2012). Video available at https://www.youtube.com/watch?v=0_a9RP0J7PA 4 minutes 10 seconds.

8. Donella H. Meadows, Dennis L. Meadows, Jorgen Randers, and William W. Behrens III, *The Limits to Growth* (New York: New American Library, 1972), 56–58.
9. Bernard Dixon, "In Praise of Prophets," *New Scientist* 51, no. 769 (Sept. 16, 1971): 606, http://books.google.com/books?id=azwQStEZq-8C&printsec=frontcover#v=onepage&q&f=false.
10. Paul Ehrlich and Anne Ehrlich, *The End of Affluence* (Riverside, MA: Rivercity Press, 1974), 49.
11. "Ecology, the New Mass Movement," *Life*, Jan. 1970, 22.
12. Paul Ehrlich, "An Interview with Ecologist Paul Ehrlich," *Mademoiselle*, Apr. 1970, 292.
13. Eric Eckholm, "Significant Rise in Sea Level Now Seems Certain," *New York Times*, Feb. 18, 1986, www.nytimes.com/1986/02/18/science/signifigant-rise-in-sea-level-now-seems-certain.html.
14. Philip Shabecoff, "Swifter Warming of Globe Foreseen," *New York Times*, June 11, 1986, www.nytimes.com/1986/06/11/us/swifter-warming-of-globe-foreseen.html.
15. Bill McKibben, *The End of Nature, rev. ed.* (New York: Random House, 2006), 124, 128.
16. Paul Ehrlich, *The Machinery of Nature* (New York: Simon & Schuster, 1986), 274.
17. Richard Lindzen, "Global Warming: The Origin and Nature of the Alleged Scientific Consensus," *Regulation: CATO Review of Business & Government* 15, no 2 (Spring 1992), 87–98, http://object.cato.org/sites/cato.org/files/serials/files/regulation/1992/4/v15n2-9.pdf.
18. Julian Simon, *The Ultimate Resource 2* (Princeton, NJ: Princeton University Press, 1996), 35.
19. Paul Ehrlich and Richard Harriman, *How to Be a Survivor*, 72.
20. Amory Lovins, "Amory Lovins: Energy Analyst and Environmentalist," *Mother Earth News*, November/December 1977, http://www.motherearthnews.com/renewable-energy/amory-lovins-energy-analyst-zmaz77ndzgoe.aspx.
21. Bill McKibben, "A Special Moment in History: Part Three," *Atlantic Online*, May 1998, http://www.theatlantic.com/past/docs/issues/98may/special3.htm.
22. ———, *Eaarth: Making a Life on a Tough New Planet* (New York: Times Books, 2010), 184.
23. Anis Shivani, "Facing Cold, Hard Truths About Global Warming," *Boston Globe*, May 30, 2010, http://www.boston.com/ae/books/articles/2010/05/30/ facing_cold_hard_truths_about_global_warming/.

24. Robert Bradley Jr., "Howlin' Wolf: Paul Ehrlich on Energy (Part I)," *MasterResources*, March 13, 2010, http://www.masterresource.org/2010/03/howlin-wolf-paul-ehrlich-on-energy-part-i-demeaning-julian-simon-energy-as-desecrator-doom-from-depletion/.
25. BP, Statistical Review of World Energy 2013.
26. Ibid.
27. Amory Lovins, "Energy Strategy: The Road Not Taken?," Foreign Affairs, October 1967, www.foreignaffairs.com/articles/26604/amory-b-lovins/energy-strategy-the-road-not-taken.
28. BP, Statistical Review of World Energy 2013.
29. BP, Statistical Review of World Energy, Historical data workbook," accessed July 10, 2014, http://www.bp.com/en/global/corporate/about-bp/energy-economics/statistical-review-of-world-energy.html.
30. World Bank, World Development Indicators (WDI) Online Data, Apr. 2014, http://data.worldbank.org/data-catalog/world-development-indicators.
31. Ibid.
32. Jimmy Carter, "The President's Proposed Energy Policy," Apr. 18, 1977, Vital Speeches of the Day 43, no. 14, May 1, 1977, www.pbs.org/wgbh/americanexperience/features/primary-resources/carter-energy.
33. Richard Heinberg, The Party's Over: Oil, War and the Fate of Industrial Societies (Gabriola Island, BC: New Society, 2005), 85.
34. Timothy R. Klett et al., "Assessment of Potential Additions to Conventional Oil and Gas Resources of the World (Outside the United States) from Reserve Growth," U.S. Geological Survey Fact Sheet, 2012, p. 2; Inderscience, "How Much Oil Have We Used?" ScienceDaily, www.sciencedaily.com/releases/2009/05/090507072830.htm.
35. R.T.H. Collis, "Weather and World Food," Bulletin of the American Meteorological Society 56, no. 10 (1975): 1078–83, http://journals.ametsoc.org/doi/abs/10.1175/1520-0477%281975%29056%3C1078%3AWAWF%3E2.0.CO%3B2.
36. John Gribbin, "Cause and Effects of Global Cooling," Nature 254 (1975), 14, doi: 10.1038/254014a0.
37. Shabecoff, "Swifter Warming of Globe Foreseen."
38. National Aeronautics and Space Administration, Goddard Institute for Space Studies, "Combined Land-Surface Air and Sea-Surface Water Temperature Anomalies (Land-Ocean Temperature Index)," table data, global-mean monthly, June 2014, http://data.giss.nasa.gov/gistemp/graphs_v3/Fig.A2.gif.
39. Bill McKibben, *The End of Nature,* 128.

40. Indur M. Goklany, "Weather and Safety: The Amazing Decline in Deaths from Extreme Weather in an Era of Global Warming, 1900–2010," Reason Foundation, Policy Study 393, Sept. 2011, http://reason.org/files/deaths_from_extreme_weather_1900_2010.pdf.
41. Ibid.
42. Intergovernmental Panel on Climate Change, "13.3.3.3 Implications of Regime Stringency: Linking Goals, Participation, and Timing," IPCC Fourth Assessment Report: Climate Change 2007, 2007, www.ipcc.ch/publications_and_data/ar4/wg3/en/ch13-ens13-3-3-3.html.
43. Intergovernmental Panel on Climate Change, "Potential of Renewable Energy Outlined in Report of Intergovernmental Panel on Climate Change," press release, May 9, 2011, http://srren.ipcc-wg3.de/press/content/potential-of-renewable-energy-outlined-report-by-the-intergovernmental-panel-on-climate-change.
44. Kirsten Gibson, "Rokita Holds Town Hall in Lebanon." Purdue Exponent, Aug. 26, 2013, www.purdueexponent.org/city_state/article_bd15b4a3-ce01-55b8-a972-27109a6eb44e.html.
45. David M. Graber, "Mother Nature as a Hothouse Flower: 'The End of Nature' by Bill McKibben," *Los Angeles Times*, Oct. 22, 1989, http://articles.latimes.com/1989-10-22/books/bk-726_1_bill-mckibben/2.
46. Bill McKibben, *The End of Nature*, 162.
47. Ibid., 183.
48. McKibben, *Eaarth: Making a Life on a Tough New Planet*, xii.
49. Haripriya Rangan, *Of Myths and Movements: Rewriting Chipko into Himalayan History* (New York: Verso, 2000), 42.

CHAPTER 2:
THE ENERGY CHALLENGE

1. U.S. Energy Information Administration, International Energy Outlook 2013, July 2013, www.eia.gov/forecasts/ieo.
2. Kathryn Hall, "Kathryn's Story," Power Up Gambia, www.powerupgambia.org/about/story.
3. "Calories Burned by Occupation," CalorieLab, http://calorielab.com/burned/?mo=se&gr=11&ti=Occupation&wt=150&un=lb&kg=68.
4. Kathleen M. Zelman, "The Olympic Diet of Michael Phelps," *WebMD Health News*, August 13, 2008, www.webmd.com/diet/news/20080813/the-olympic-diet-of-michael-phelps?printing=true.

5. World Bank, World Development Indicators (WDI) Online Data, http://data.worldbank.org/data-catalog/world-development-indicators.
6. Milton Friedman and Rose Friedman, *Free to Choose: A Personal Statement* (San Diego: Harcourt Brace, 1980), 148.
7. International Energy Agency, "Energy Poverty," www.iea.org/topics/energypoverty.
8. Ibid.
9. World Bank, World Development Indicators (WDI) Online Data, April 2014, http://data.worldbank.org/data-catalog/world-development-indicators.
10. *Saturday Night Live,* Jimmy Fallon, NBC, Dec. 9, 2000.
11. Food and Agriculture Organization, Regional Office for Europe, "Inventory of Hazelnut Research, Germplasm and Reference," (accessed July 17, 2014), http://www.fao.org/docrep/003/x4484e/x4484e03.htm.
12. BP, Statistical Review of World Energy 2013, www.bp.com/en/global/corporate/about-bp/energy-economics/statistical-review-of-world-energy-2013.html.
13. Ibid.
14. American Enterprise Institute, "The Myth of Green Energy Jobs," http://www.aei.org/outlook/energy-and-the-environment/the-myth-of-green-energy-jobs-the-european-experience/.
15. Department of Energy, Office of Energy Efficiency and Renewable Energy, "Dye-Sensitized Solar Cells," http://energy.gov/eere/sunshot/dye-sensitized-solar-cells; Department of Energy, Office of Energy Efficiency and Renewable Energy, "Photovoltaic Cell Material Basics," August 19, 2013, http://energy.gov/eere/energybasics/articles/photovoltaic-cell-material-basics.
16. GE Power and Water, Renewable Energy, "1.5 MW Wind Turbine Series," 2010, http://site.ge-energy.com/prod_serv/products/wind_turbines/en/downloads/GEA14954C15-MW-Broch.pdf.
17. Stephen Markley, "The Rumpus Interview with Bill McKibben," *The Rumpus,* December 10, 2012, accessed July 10, 2014, http://therumpus.net/2012/12/the-rumpus-interview-with-bill-mckibben/.
18. Kiley Kroh, "Germany Sets New Record, Generating 74 Percent of Power Needs from Renewable Energy," thinkprogress.org, May 13, 2014, http://thinkprogress.org/climate/2014/05/13/3436923/germany-energy-records/.

19. I use daily data for solar and wind electricity production and monthly data for total electricity production because each was the most specific data I could obtain from the German government.
20. Reuters, "Germany Plans to Build, Revamp 84 Power Plants-BDEW," April 23, 2012, http://www.reuters.com/article/2012/04/23/germany-energy-bdew-idUSF9E7J100V20120423.
21. Bruno Burger, "Electricity Production from Solar and Wind in Germany," Fraunhofer Institute for Solar Energy Systems, presentation, Freiburg, Germany, May 26, 2014, www.ise.fraunhofer.de/en/downloads-englisch/pdf-files-englisch/data-nivc-/electricity-production-from-solar-and-wind-in-germany-2014.pdf.
22. *The Great Global Warming Swindle*, directed by Martin Durkin, Mar. 8, 2007, WAGtv, http://greatglobalwarmingswindle.com.
23. Ibid.
24. "Wood: The Fuel of the Future," *Economist*, Apr. 6, 2013, www.economist.com/news/business/21575771-environmental-lunacy-europe-fuel-future.
25. "State of Food Insecurity in the World," Food and Agriculture Organization of the United Nations, 2011, p. 11, www.fao.org/docrep/014/i2330e/i2330e03.pdf.
26. BP, Statistical Review of World Energy 2013.
27. Ibid.
28. Cedric Philibert and Carlos Gasco, *Technology Roadmap: Hydropower* (Paris: International Energy Agency, 2012), www.iea.org/publications/freepublications/publication/2012_Hydropower_Roadmap.pdf, 18.
29. Ibid., 5.
30. "About," *Years of Living Dangerously*, www.sho.com/sho/years-of-living-dangerously/about.
31. Reed McManus, "Down Come the Dams," *Sierra*, May–June 1998, www.sierraclub.org/sierra/199805/lol.asp.
32. *International Atomic Energy Agency*, "Nuclear Power Advantages," www.iaea.org/Publications/Booklets/Development/devnine.html (accessed June 22, 2014).
33. World Nuclear Association, "Safety of Nuclear Power Reactors," Apr. 2014, www.world-nuclear.org/info/Safety-and-Security/Safety-of-Plants/Safety-of-Nuclear-Power-Reactors.
34. World Nuclear Association, "Nuclear Radiation and Health Effects," Feb. 2014, www.world-nuclear.org/info/Safety-and-Security/Radiation-and-Health/Nuclear-Radiation-and-Health-Effects.

CHAPTER 3:
THE GREATEST ENERGY TECHNOLOGY OF ALL TIME

1. Vaclav Smil, *Energy in World History* (Boulder, CO: Westview Press, 1994), 116.
2. International Energy Agency, *World Energy Outlook 2013,* November 12, 2013, www.worldenergyoutlook.org/publications/weo-2013.
3. World Bank, World Development Indicators (WDI) Online Data, April 2014, http://data.worldbank.org/data-catalog/world-development-indicators.
4. BP, Statistical Review of World Energy 2013, www.bp.com/en/global/corporate/about-bp/energy-economics/statistical-review-of-world-energy-2013.html.
5. World Bank, World Development Indicators (WDI) Online Data.
6. Smil, *Energy in World History,* 160–66.
7. "Shenhua Coal to Liquids Plant, China," *Hydrocarbons Technology,* www.hydrocarbons-technology.com/projects/shenhua (accessed June 23, 2014).
8. BP, Statistical Review of World Energy 2013.
9. Ibid.
10. Hobart King, "Methane Hydrates," *Geology.com,* http://geology.com/articles/methane-hydrates (accessed June 23, 2014).
11. U.S. Energy Information Administration, "Annual Energy Outlook 2014," Table A2: Energy Consumption by Sector and Source, May 2014, www.eia.gov/forecasts/aeo/pdf/tbla2.pdf.
12. Tesla Motors, "Battery: Increasing Energy Density Means Increasing Range," 2014, www.teslamotors.com/roadster/technology/battery.
13. U.S. Energy Information Administration, "How Much Gasoline Does the United States Consume?" May 13, 2014, www.eia.gov/tools/faqs/faq.cfm?id=23&t=10; BP, *Statistical Review of World Energy 2013;* World Bank, World Development Indicators (WDI) Online Data.
14. American Chemical Society, "Development of the Pennsylvania Oil Industry," National Historic Chemical Landmarks Program, 2009 www.acs.org/content/acs/en/education/whatischemistry/landmarks/pennsylvaniaoilindustry.html.
15. Daniel Yergin, *The Prize: The Epic Quest for Oil, Money & Power* (New York: Simon & Schuster, 2008), 11.
16. National Resources Defense Council, "Stopping the Keystone Pipeline," www.nrdc.org/energy/keystone-pipeline (accessed June 23, 2014).

17. William Stanley Jevons, *The Coal Question: An Inquiry Concerning the Progress of the Nation, and the Probable Exhaustion of Our Coal Mines* (London: Macmillan, 1865), vii.
18. Ibid., viii.
19. Anonymous, *Times (London),* Apr. 19, 1866, repr. in Sandra Peart, ed., *W. S. Jevons: Critical Responses,* 4 vols. (New York: Routledge, 2003), 1: 196.
20. Ibid.
21. Paul Ehrlich and Anne Ehrlich, *The Population Bomb* (New York: Sierra Club/Ballantine, 1968), xi.
22. Julian Simon, *The Ultimate Resource 2* (Princeton: Princeton University Press, 1981), 85.
23. Robert Sirico, *Defending the Free Market: The Moral Case for a Free Economy* (Washington, DC: Regnery, 2012), 164.
24. "World Population," United States Census Bureau, Dec. 2013, www.census.gov/population/international/data/worldpop/table_population.php.
25. World Bank, World Development Indicators (WDI) Online Data.
26. Matt Ridley, *The Rational Optimist: How Prosperity Evolves* (New York: Harper, 2010), 157.
27. William Crookes, "Presidential Address to the British Association for the Advancement of Science," *Chemical News* 78 (1898): 125.
28. University of York, "Ammonia," The Essential Chemical Industry, Jan. 2, 2014, www.essentialchemicalindustry.org/chemicals/ammonia.html.
29. Indur M. Goklany, "Humanity Unbound: Fossil Fuels Saved Humanity from Nature and Nature from Humanity," Cato Institute, Policy Analysis 715, Dec. 12, 2012, 3, www.cato.org/sites/cato.org/files/pubs/pdf/pa715.pdf.
30. BP, Statistical Review of World Energy 2013.
31. John Kerry, "Remarks on Climate Change," U.S. Department of State, speech, Jakarta, Indonesia, Feb. 16, 2014, www.state.gov/secretary/remarks/2014/02/221704.htm.

CHAPTER 4:
THE GREENHOUSE EFFECT AND THE FERTILIZER EFFECT

1. Richard Lindzen, "Some Coolness Concerning Global Warming," *Bulletin of the American Meteorological Society* 71, no. 3, Mar. 1990, http:

//eaps.mit.edu/faculty/lindzen/cooglobwrm.pdf; Patrick J. Michaels, *Sound and Fury: The Science and Politics of Global Warming* (Washington, DC: Cato Institute, 1992).

2. Lawrence A. Baker et al., "Urbanization and Warming of Phoenix (Arizona, USA): Impacts, Feedbacks and Mitigation," *Urban Ecosystems* 6, no. 3 (Sept. 2002): 183–203, doi: 10.1023/A:1026101528700.
3. "20,000 Killed by Earthquake: Toll Is Growing, Bodies Float Down Ganges to the Sea," *Border Cities Star*, January 20, 1934, http://news.google.com/newspapers?id=3Q4_AAAAIBAJ&sjid=aU4MAAAAIBAJ&dq=earthquake&pg=1690%2C3778979.
4. "100 Are Injured, Property Damage Exceeds $1,000,000: Tornado Strikes Three States, Bitter Cold in North Area," *Ames Daily Tribune*, February 26, 1934, http://www.newspapers.com/image/?spot=605947.
5. "Death's Toll Mounts to 60 in U.S. Storms," *Montreal Gazette*, February 27, 1934, http://news.google.com/newspapers?id=O0swAAAAIBAJ&sjid=bKgFAAAAIBAJ&dq=bering%20sea%20ice&pg=6065%2C3419263.
6. "1,500 Japanese Die in Hakodate Fire; 200,000 Homeless: Largest City North of Tokyo Is in Ruins and Mayor Says It Is 'a Living Hell,'" *New York Times*, March 22, 1934, http://timesmachine.nytimes.com/timesmachine/1934/03/22/94504668.html.
7. "Where Tidal Wave Ruined Norway Fishing Towns," *Lewiston Morning News*, April 26, 1934, http://news.google.com/newspapers?id=wKJfAAAAIBAJ&sjid=HDIMAAAAIBAJ&dq=tidal%20wave&pg=1854%2C6770134.
8. "Antarctic Heat Wave: Explorers Puzzled but Pleased," *West Australian*, Perth, Australia, June 7, 1934, http://trove.nla.gov.au/ndp/del/article/33242855.
9. "7 Lives Lost as Tropical Storm Whips Louisiana: Hurricane Moves Far Inland Before Blowing Out Its Wrath in Squalls," *Tuscaloosa News*, June 18, 1934, http://news.google.com/newspapers?id=n34AAAAIBAJ&sjid=7EsMAAAAIBAJ&pg=5144%2C2222085.
10. "Widely Separated Regions of the Globe Feel Heavy Quake," *St. Petersburg Times*, June 21, 1934, http://news.google.com/newspapers?id=Fy9PAAAAIBAJ&sjid=e04DAAAAIBAJ&dq=earthquake&pg=4691%2C1046632.
11. "Earth Growing Warmer: What Swiss Glaciers Reveal," *Courier-Mail*, June 22, 1934, http://trove.nla.gov.au/ndp/del/article/36712632?searchTerm=%22climate%20change%22&searchLimits=.

12. "Death, Suffering over Wide Area in China Drouth," *Southeast Missourian*, July 17, 1934, http://news.google.com/newspapers?id=wawoAAAAIBAJ&sjid=xdIEAAAAIBAJ&pg=6058,6240793&dq=heat-wave+china&hl=en.
13. Wladislaw Besterman, "Toll of Flood at High Figure: Over 100 Bodies Recovered and 500 Persons Missing in Southern Poland," *Waycross Journal-Herald*, July 18, 1934, http://news.google.com/newspapers?id=UGFaAAAAIBAJ&sjid=8EwNAAAAIBAJ&dq=heat%20nebraska& pg=4328%2C5922735.
14. "Cuban Malaria Increases: Thousands Become Ill in Usual Seasonal Spread of Disease," *New York Times*, July 23, 1934, http://timesmachine.nytimes.com/timesmachine/1934/07/23/93634956.html.
15. "Mid-West Hopes for Relief from Heat; 602 Killed," *Berkeley Daily Gazette*, Berkeley, California, July 25, 1934, http://news.google.com/newspapers?id=MDcyAAAAIBAJ&sjid=FOMFAAAAIBAJ&pg=227 6%2C1837317.
16. "Famine Faces 5,000,000 in Drouth [Area," *Pittsburgh Press*, August 14, 1934, http://news.google.com/newspapers?nid=djft3U1LymYC&dat=19340814&printsec=frontpage.
17. "Rumanians Are Alarmed by Epidemic of Cholera," *New York Times*, September 9, 1934, http://timesmachine.nytimes.com/timesmachine/1934/09/10/95056316.html.
18. James E. Hansen, "Twenty Years Later: Tipping Points Near on Global Warming," *Huffington Post*, June 23, 2008, www.huffingtonpost.com/dr-james-hansen/twenty-years-later-tippin_b_108766.html.
19. Lindzen, "Some Coolness Concerning Global Warming."
20. Qiancheng Ma, "Greenhouse Gases: Refining the Role of Carbon Dioxide," Science Briefs, National Aeronautics and Space Administration, Goddard Institute for Space Studies, Mar. 1998, www.giss.nasa.gov/research/briefs/ma_01.
21. Patrick Lynch, "Paleoclimate Record Points Toward Potential Rapid Climate Changes," National Aeronautics and Space Administration, Goddard Institute for Space Studies, Research News, Dec. 8, 2011, www.giss.nasa.gov/research/news/20111208.
22. David M. Etheridge et al., "Natural and Anthropogenic Changes in Atmospheric CO_2 over the Last 1,000 Years from Air in Antarctic Ice and Firn," *Journal of Geophysical Research* 101, no. D2 (Feb. 20, 1996): 4115–28, doi: 10.1029/95JD03410.
23. Gunnar Myhre, et al., "New Estimates of Radiative Forcing Due to Well-Mixed Greenhouse Gases," *Geophysical Research Letters* 25 (1998). No. 14, pp. 2715–18.

24. John Kerry, "Remarks on Climate Change," U.S. Department of State, Speech, Jakarta, Indonesia, Feb. 16, 2014, www.state.gov/secretary/remarks/2014/02/221704.htm.
25. "The Keeling Curve: A Daily Record of Atmospheric Carbon Dioxide from Scripps Institution of Oceanography at UC San Diego," June 8, 2014, http://keelingcurve.ucsd.edu.
26. National Aeronautics and Space Administration, Goddard Institute for Space Studies, "Global Land-Ocean Temperature Index," GISS Surface Temperature Plots, May 12, 2014, http://data.giss.nasa.gov/gistemp/graphs_v3/Fig.A2.gif.
27. Davis Guggenheim, director, *An Inconvenient Truth* (Beverly Hills, CA: Lawrence Bender Production), 2006.
28. Svante Arrhenius, *Worlds in the Making: The Evolution of the Universe* (New York: Harper & Brothers, 1908), 63.
29. "Global Warming's Denier Elite," *Rolling Stone Online*, September 12, 2013, www.rollingstone.com/politics/news/global-warmings-denier-elite-20130912.
30. National Aeronautics and Space Administration," Consensus: 97 Percent of Climate Scientists Agree," http://climate.nasa.gov/scientific-consensus (accessed June 7, 2014).
31. Kerry, "Remarks on Climate Change," Feb. 16, 2014.
32. President Barack Obama, Twitter post, May 16, 2013, 12:48 p.m., https://twitter.com/BarackObama/status/335089477296988160.
33. John Cook et al., "Quantifying the Consensus on Anthropogenic Global Warming in the Scientific Literature," *Environmental Research Letters* 8, May 15, 2013, doi:10.1088/1748-9326/8/2/024024.
34. David Friedman, "A Climate Falsehood You Can Check for Yourself," *Ideas*, Feb. 26, 2014, http://daviddfriedman.blogspot.com/2014/02/a-climate-falsehood-you-can-check-for.html.
35. Cook et al., "Quantifying the Consensus."
36. Ibid.
37. Andrew (computer analyst), "Cook's 97 Percent Consensus Study Falsely Classifies Scientists' Papers According to the Scientists That Published Them," *Popular Technology*, May 21, 2013, www.populartechnology.net/2013/05/97-study-falsely-classifies-scientists.html.
38. Ibid.
39. Ibid.
40. Ibid.
41. Jonathan Schell, "Our Fragile Earth," *Discover*, Oct. 1989, 45–48.

42. Paul Ehrlich and Anne Ehrlich, *Betrayal of Science and Reason* (Washington, DC: Island Press, 1996), 207.
43. Craig D. Idso, "The State of Earth's Terrestrial Biosphere: How Is It Responding to Rising Atmospheric CO_2 and Warmer Temperatures?" Center for the Study of Carbon Dioxide and Global Change, Dec. 5, 2012, www.co2science.org/education/reports/greening/TheStateofEarthsTerrestrialBiosphere.pdf.
44. Craig D. Idso, "The Positive Externalities of Carbon Dioxide: Estimating the Monetary Benefits of Rising Atmospheric CO_2 Concentrations on Global Food Production," Center for the Study of Carbon Dioxide and Global Change, Oct. 21, 2013, www.co2science.org/education/reports/co2benefits/co2benefits.php.
45. Monte Hieb, "Climate and the Carboniferous Period," *Plant Fossils of West Virginia,* Mar. 21, 2009, www.geocraft.com/WVFossils/Carboniferous_climate.html.

CHAPTER 5:
THE ENERGY EFFECT AND CLIMATE MASTERY

1. EM-DAT: OFDA/CRED International Disaster Database, U.S. Office of Foreign Disaster Assistance and Centre for Research on the Epidemiology of Disasters, Université catholique de Louvain, Brussels, Belgium, 2014, home, www.emdat.be.
2. Ibid.
3. Ibid.
4. Ibid.
5. National Oceanic and Atmospheric Administration, National Climatic Data Center, "U.S. Tornado Climatology," www.ncdc.noaa.gov/climate-information/extreme-events/us-tornado-climatology (accessed June 25, 2014); National Oceanic and Atmospheric Administration, National Climatic Data Center, "Hurricanes," www.ncdc.noaa.gov/oa/climate/severeweather/hurricanes.html (accessed June 25, 2014).
6. Indur M. Goklany, "Weather and Safety: The Amazing Decline in Deaths from Extreme Weather in an Era of Global Warming, 1900–2010," Reason Foundation, Policy Study 393, Sept. 2011, 3, http://reason.org/files/deaths_from_extreme_weather_1900_2010.pdf.
7. Indur M. Goklany, "*Humanity Unbound: Fossil Fuels Saved Humanity from Nature and Nature from Humanity,*" Cato Institute, Policy Analysis 715, Dec. 12, 2012, 3, www.cato.org/sites/cato.org/files/pubs/pdf/pa715.pdf.

8. EM-DAT database.
9. John Kerry, "Remarks on Climate Change," U.S. Department of State, speech, Jakarta, Indonesia, Feb. 16, 2014, www.state.gov/secretary/remarks/2014/02/221704.htm.
10. Bill McKibben, "Global Warming's Terrifying New Math," *Rolling Stone,* July 19, 2012, www.rollingstone.com/politics/news/global-warmings-terrifying-new-math-20120719.
11. WorldLifeExpectancy.com, "USA Life Expectancy," www.worldlifeexpectancy.com/usa/life-expectancy (accessed June 25, 2014).
12. Joanna Zelman, "50 Million Environmental Refugees by 2020, Experts Predict," *Huffington Post,* May 25, 2011, www.huffingtonpost.com/2011/02/22/environmental-refugees-50_n_826488.html.
13. IPCC, 2013: Summary for Policymakers. In: Climate Change 2013: The Physical Science Basis. Contribution of Working Group I to the Fifth Assessment Report of the Intergovernmental Panel on Climate Change [T. F. Stocker, et al., eds.]. Cambridge University Press, Cambridge, UK, and New York, NY, p. 23, doi:10.1017/CBO978110 74153 24.004, http://www.climatechange2013.org/images/report/WG1AR5_SPM_FINAL.pdf.
14. Vanessa McKinney, "Sea Level Rise and the Future of the Netherlands," ICE Case Studies 212, May 2007, www1.american.edu/ted/ice/dutch-sea.htm.
15. J. W. de Zeeuw, trans., International Peat Society Commission VIII, "Peat and the Dutch Golden Age: The Historical Meaning of Energy Attainability," *A.A.G. Bijdragen* 21 (1978): 5–6.
16. Richard Tol and Andreas Langen, "A Concise History of Dutch River Floods," *Climate Change* 46, no. 3 (Aug. 2000): 359–60.
17. Ibid., 361.
18. Albert Gore Jr., *Earth in the Balance: Ecology and the Human Spirit* (Boston: Houghton Mifflin, 1992; New York: Rodale, 2006), 326; citation to 2006 ed.
19. Petr Beckmann, *The Health Hazards of Not Going Nuclear* (Boulder, CO: Golem Press, 1979).
20. McKibben, "Global Warming's Terrifying New Math"; James E. Hansen, "Twenty Years Later: Tipping Points Near on Global Warming," *Huffington Post,* June 23, 2008, www.huffingtonpost.com/dr-james-hansen/twenty-years-later-tippin_b_108766.html.
21. Fred Krupp, "Climate Change Opportunity," *Wall Street Journal,* Apr. 8, 2008, http://online.wsj.com/news/articles/SB120761565455196769.

22. BP, Statistical Review of World Energy 2013, www.bp.com/en/global/corporate/about-bp/energy-economics/statistical-review-of-world-energy-2013.html.
23. Goklany, "Weather and Safety," 3.
24. Ayn Rand, *Atlas Shrugged* (New York: Random House, 1957), 146–47.

CHAPTER 6:
IMPROVING OUR ENVIRONMENT

1. "The ocean holds 97 percent of the Earth's water; the remaining three percent is freshwater found in glaciers and ice, below the ground, or in river or lakes," National Oceanic and Atmosphere Administration, National Ocean Service, "Where Is All of the Earth's Water?" Jan. 11, 2013, http://oceanservice.noaa.gov/facts/wherewater.html; U.S. Geological Survey, USGS Water Science School, "The World's Water," Mar. 17, 2014, http://water.usgs.gov/edu/earthwherewater.html.
2. U.S. Environmental Protection Agency, "Drinking Water Contaminants," June 3, 2014, http://water.epa.gov/drink/contaminants/index.cfm; U.S. Environmental Protection Agency, "Drinking Water Treatment," 2009, http://water.epa.gov/lawsregs/guidance/sdwa/upload/2009_08_28_sdwa_fs_30ann_treatment_web.pdf.
3. "Water-Related Diseases: Information Sheets," World Health Organization, accessed July 10, 2014, http://www.who.int/water_sanitation_health/diseases/diseasefact/.
4. Sara E. Rollins, Sean M. Rollins, and Edward T. Ryan, "*Yersinia pestis* and the Plague," *American Journal of Clinical Pathology* 119 (2003), S78, doi:10.1309/DQM93R8QNQWBFYU8; David Leon, "Smallpox and Global Governance," *Global Policy*, www.globalpolicyjournal.com/brookings-audit/smallpox-and-global-governance; Richard Carter and Kamini N. Mendis, "Evolutionary and Historical Aspects of the Burden of Malaria," *Clinical Microbiology Reviews* 15, no. 4 (Oct. 2002): table 3, doi: 10.1128/CMR.15.4.564-594.2002.
5. Paul Reiter, Memorandum for the British House of Lords Select Committee on Economic Affairs, Mar. 31, 2005, www.publications.parliament.uk/pa/ld200506/ldselect/ldeconaf/12/12we21.htm.
6. World Health Organization, "Cholera 2012," *Weekly Epidemiological Record, no. 31,* Aug. 2, 2013, p. 321, www.who.int/wer/2013/wer8831.pdf?ua=1.

7. Amy Standen, "Here, Drink a Nice Glass of Sparkling Clear Wastewater," National Public Radio, Nov. 7, 2013, www.npr.org/2013/11/07/243711364/here-drink-a-nice-glass-of-sparkling-clean-wastewater.

CHAPTER 7: REDUCING RISKS AND SIDE EFFECTS

1. Flora Felsovalyi, Bennett Jap, Anthony Robilotto, and Gary Tong, "Delayed Effects of Hydrofluoric Acid Burn," *Cornell University Library eCommons*, Jan. 7, 2001, http://hdl.handle.net/1813/262.
2. Robert Bryce, *Smaller, Faster, Lighter, Denser, Cheaper* (New York: Perseus Book Groups, 2014), 192.
3. Simon Parry and Ed Douglas, "In China, the True Cost of Britain's Clean, Green Wind Power Experiment: Pollution on a Disastrous Scale," *London Daily Mail*, Jan. 26, 2011, www.dailymail.co.uk/home/moslive/article-1350811/In-China-true-cost-Britains-clean-green-wind-power-experiment-Pollution-disastrous-scale.html.
4. Stephen Mosley, "Public Perception of Smoke Pollution in Victorian Manchester," in *Technologies of Landscape: From Reaping to Recycling*, ed. David E. Nye (Amherst, MA: University of Massachusetts Press, 1999), 163.
5. Ibid.
6. Steve Tracton, "The Killer London Smog Event of December, 1952: A Reminder of Deadly Smog Events in U.S.," *Washington Post*, blogs, Dec. 19, 2012, www.washingtonpost.com/blogs/capital-weather-gang/post/the-killer-london-smog-event-of-december-1952-a-reminder-of-deadly-smog-events-in-us/2012/12/19/152c66bc-498e-11e2-b6f0-e851e741d196_blog.html.
7. BP Global, "Deepwater Horizon Accident and Response," Gulf of Mexico Restoration, Dec. 13, 2013, www.bp.com/en/global/corporate/gulf-of-mexico-restoration/deepwater-horizon-accident-and-response.html.
8. Daniel Yergin, *The Prize: The Epic Quest for Oil, Money and Power* (New York: Simon & Schuster, 2008), 34.
9. U.S. Department of Labor, Bureau of Labor Statistics, Census of Fatal Occupational Injuries, Fatal Injury Rate by Industry 2012, accessed May 4, 2014, www.bls.gov/iif/oshwc/cfoi/cfoi_rates_2012hb.pdf, 2.
10. *Gasland*, directed by Josh Fox (New York: HBO Documentary Films and International WOW Company, 2010), http://one.gaslandthemovie

.com/home; *Gasland 2*, directed by Josh Fox (New York: HBO Documentary Films and International WOW Company, Apr. 21, 2013), www.gaslandthemovie.com.

11. Colorado Oil & Gas Association, "COGA's the Truth About 'GasLand,'" June 17, 2011, http://www.coga.org/FactSheets/FactSheetGasLand.pdf.
12. Leslie A. DeSimone, Pixie A. Hamilton, Robert J. Gilliom, "Quality of Water from Domestic Wells in Principal Aquifers of the United States, 1991–2004, Overview of Major Findings," National Water-Quality Assessment Program, Circular 1332, U.S. Department of the Interior, 2009, pp. 17–18, http://pubs.usgs.gov/circ/circ1332/includes/circ1332.pdf.
13. "EPA Jackson 'Not Aware of Any Proven Case Where the Fracking Process Itself Has Affected Water,'" Senate Committee on Environment & Public Works, press release, May 24, 2011, www.epw.senate.gov/public/index.cfm?FuseAction=Minority.PressReleases&ContentRecord_id=23eb85dd-802a-23ad-43f9-da281b2cd287.
14. Duncan Graham-Rowe, "Lifestyle: When Allergies Go West," *Nature* 479, no. 7374 (Nov. 24, 2011): S2–S4, doi: 10.1038/479S2a.
15. Heinrich Duhme et al., "Asthma and Allergies Among Children in West and East Germany: A Comparison Between Münster and Greifswald Using the ISAAC Phase I Protocol," International Study of Asthma and Allergies in Childhood, *European Respiratory Journal* 11, no. 4 (Apr. 1988): 840–47, www.ncbi.nlm.nih.gov/pubmed/9623686.
16. Ibid.
17. Willie Soon and Paul Driessen, "The Myth of Killer Mercury: Panicking People About Fish Is No Way to Protect Public Health," *Wall Street Journal,* May 25, 2011, http://online.wsj.com/article/SB10001424052748703421204576329420414284558.html; Willie Soon, "A Scientific Critique of the Environmental Protection Agency's NESHAP Proposed Rule," March 16, 2011, 29, www.cfact.org/pdf/Scientific_Critique_of_EPA_MercuryRule062011.pdf.
18. P. W. Davidson et al., "Fish Consumption and Prenatal Methylmercury Exposure: Cognitive and Behavioral Outcomes in the Main Cohort at 17 Years from the Seychelles Child Development Study," *Neurotoxicology* 32, no. 6 (December 2011): 711–17, www.ncbi.nlm.nih.gov/pmc/articles/PMC3208775/pdf/nihms327840.pdf.
19. "Paracelsus," Toxipedia, Nov. 12, 2013, www.toxipedia.org/display/toxipedia/Paracelsus.

20. Sierra Club, "Beyond Natural Gas," http://content.sierraclub.org/naturalgas (accessed May 8, 2014).
21. N. R. Warpinski, J. Du, and U. Zimmer, "Measurements of Hydraulic-Fracture-Induced Seismicity," Society of Petroleum Engineers, SPE 151597, 2012, http://www.energy4me.org/hydraulicfracturing/wp-content/uploads/2013/08/SPE-151597-MS-P1.pdf.
22. Ibid.
23. American Petroleum Institute, "The Facts About Hydraulic Fracturing and Seismic Activity," 2014, www.api.org/~/media/Files/Policy/Hydraulic_Fracturing/HF-and-Seismic-Activity-Report-v2.pdf.
24. Pierre Desrochers and Hiroko Shimizu, "Innovation and the Greening of Alberta's Oil Sands," Montreal Economic Institute, Oct. 2012, p. 16, www.iedm.org/files/cahier1012_en.pdf.
25. World Bank, World Development Indicators (WDI) Online Data.
26. World Bank, World Development Indicators (WDI) Online Data; World Health Organization, "Global Tuberculosis Report 2013," 2013, www.who.int/tb/publications/global_report/en.
27. World Bank, World Development Indicators (WDI) Online Data.
28. Ibid.

CHAPTER 8:
FOSSIL FUELS, SUSTAINABILITY, AND THE FUTURE

1. David Miliband, "Speech by David Miliband to Congress 2006," Trades Union Congress, Sept. 12, 2006, www.tuc.org.uk/about-tuc/congress/congress-2006/speech-david-miliband-congress-2006.
2. Bill McKibben, *Deep Economy: The Wealth of the Communities and the Durable Future* (New York: Times Books, 2007), 164.
3. Paul R. Ehrlich and John Holdren, *Global Ecology: Readings Toward a Rational Strategy for Man* (New York: Harcourt Brace Jovanovich, 1971), 7.
4. Paul R. Ehrlich, Anne H. Ehrlich, and John Holdren, *Human Ecology: Problems and Solutions* (San Francisco: W. H. Freeman, 1973), 279.
5. Ibid.
6. Glenn M. Ricketts, "The Roots of Sustainability," National Association of Scholars, Jan. 19, 2010, www.nas.org/articles/The_Roots_of_Sustainability.

CHAPTER 9:
WINNING THE FUTURE

1. Anis Shivani, "Facing Cold, Hard Truths About Global Warming," *Boston Globe,* May 30, 2010, www.boston.com/ae/books/articles/2010/05/30/facing_cold_hard_truths_about_global_warming.
2. Bill McKibben, "Global Warming's Terrifying New Math," *Rolling Stone,* July 19, 2012, www.rollingstone.com/politics/news/global-warmings-terrifying-new-math-20120719.
3. Joe Romm, "McKibben Must-Read: 'Global Warming's Terrifying New Math,'" *Climate Progress* blog, *Think Progress,* July 23, 2012, http://thinkprogress.org/climate/2012/07/23/565751/mckibben-must-read-global-warming039s-terrifying-new-math.
4. Jane Mayer, "Taking It to the Streets," *New Yorker,* Nov. 28, 2011, www.newyorker.com/talk/comment/2011/11/28/111128taco_talk_mayer.
5. "Review of Emerging Resources: U.S. Shale Gas and Shale Oil Plays," U.S. Energy Information Administration, July 2011, U.S. Department of Energy, www.eia.gov/analysis/studies/usshalegas/pdf/usshaleplays.pdf.
6. Steve Horn, "NY Assembly Passes Two-Year Fracking Moratorium, Senate Expected to Follow," *Huffington Post* blog, Mar. 7, 2013, www.huffingtonpost.com/steve-horn/ny-assembly-fracking-moratorium_b_2831272.html.
7. Tripp Baltz, "Court Upholds Imposing Fracking Ban in Colorado City," *Bloomberg,* Mar. 3, 2014, www.bloomberg.com/news/2014-03-03/court-upholds-imposing-fracking-ban-in-colorado-city.html.
8. "Fracking Moratorium Fails in California Despite Strong Public Support," *RT,* May 30, 2014, http://rt.com/usa/162616-california-senate-kills-fracking-ban.
9. International Energy Agency, "World Energy Outlook 2013," Nov. 12, 2013, www.worldenergyoutlook.org/publications/weo-2013.
10. David C. Scott and James A. Luppens, "Assessment of Coal Geology, Resources, and Reserve Base in the Powder River Basin, Wyoming and Montana," U.S. Geological Survey, Feb. 26, 2013, http://pubs.usgs.gov/fs/2012/3143/fs-2012-3143.pdf.
11. Patrick Rucker, "Analysis: Coal Fight Looms, Keystone-like, over U.S. Northwest," Reuters, Sept. 23, 2012, www.reuters.com/article/2012/09/23/us-coal-keystone-idUSBRE88M07F20120923.
12. Emily L., "Robert F. Kennedy, Jr: 'Coal is Crime,'" Care2, May 8, 2012, www.care2.com/causes/robert-f-kennedy-jr-coal-is-crime.html.

13. Paul Ciotti, "Fear of Fusion: What If It Works?" *Los Angeles Times*, Apr. 19, 1989, May 9, 2014, http://articles.latimes.com/1989-04-19/news/vw-2042_1_fusion-uc-berkeley-inexhaustible.
14. Ibid.
15. Rael Jean Isaac and Erich Isaac, *The Coercive Utopians: Social Deception by America's Power Players* (Washington, DC: Regnery Gateway, 1984), 7.
16. Ibid.
17. Prince Philip, foreword to *If I Were an Animal* (New York: William Morrow, 1987).
18. David M. Graber, "Mother Nature as a Hothouse Flower: 'The End of Nature' by Bill McKibben," *Los Angeles Times*, Oct. 22, 1989, http://articles.latimes.com/1989-10-22/books/bk-726_1_bill-mckibben.
19. Bryan Walsh, "Exclusive: How the Sierra Club Took Millions from the Natural Gas Industry—and Why They Stopped," *Time*, Feb. 2, 2012, http://science.time.com/2012/02/02/exclusive-how-the-sierra-club-took-millions-from-the-natural-gas-industry-and-why-they-stopped.
20. Barack Obama, "Full Text: Obama's Foreign Policy Speech," July 16, 2008, speech, Washington, DC, *Guardian* (Manchester), www.theguardian.com/world/2008/jul/16/uselections2008.barackobama.
21. Barack Obama, "Remarks to the Detroit Economic Club," May 7, 2007, speech, Detroit, MI, The American Presidency Project, www.presidency.ucsb.edu/ws/index.php?pid=77000 (accessed May 9, 2014).

INDEX

Note: Page numbers in *italics* refer to charts and graphs.